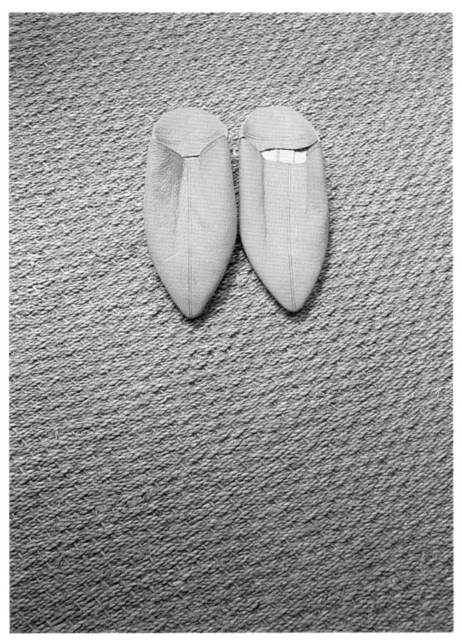

flooring

flooring

the essential source book for planning, selecting and restoring floors

Elizabeth Wilhide *with photography by* **Henry Bourne**

RYLAND
PETERS
& SMALL
LONDON NEW YORK

First published in 1997.
This revised edition published in 2005 by
Ryland Peters & Small
20–21 Jockey's Fields
London WC1R 4BW
www.rylandpeters.com

10 9 8 7 6 5 4 3 2 1

ISBN 1 84172 997 3

A catalogue record for this book is available from the British Library.

For this edition:

Designer *Sarah Fraser*
Senior editor *Clare Double*
Picture research *Emily Westlake*
Production *Gavin Bradshaw*
Art director *Gabriella Le Grazie*
Editorial director *Julia Charles*
Illustrator *John Woodcock*

Contents

Introduction

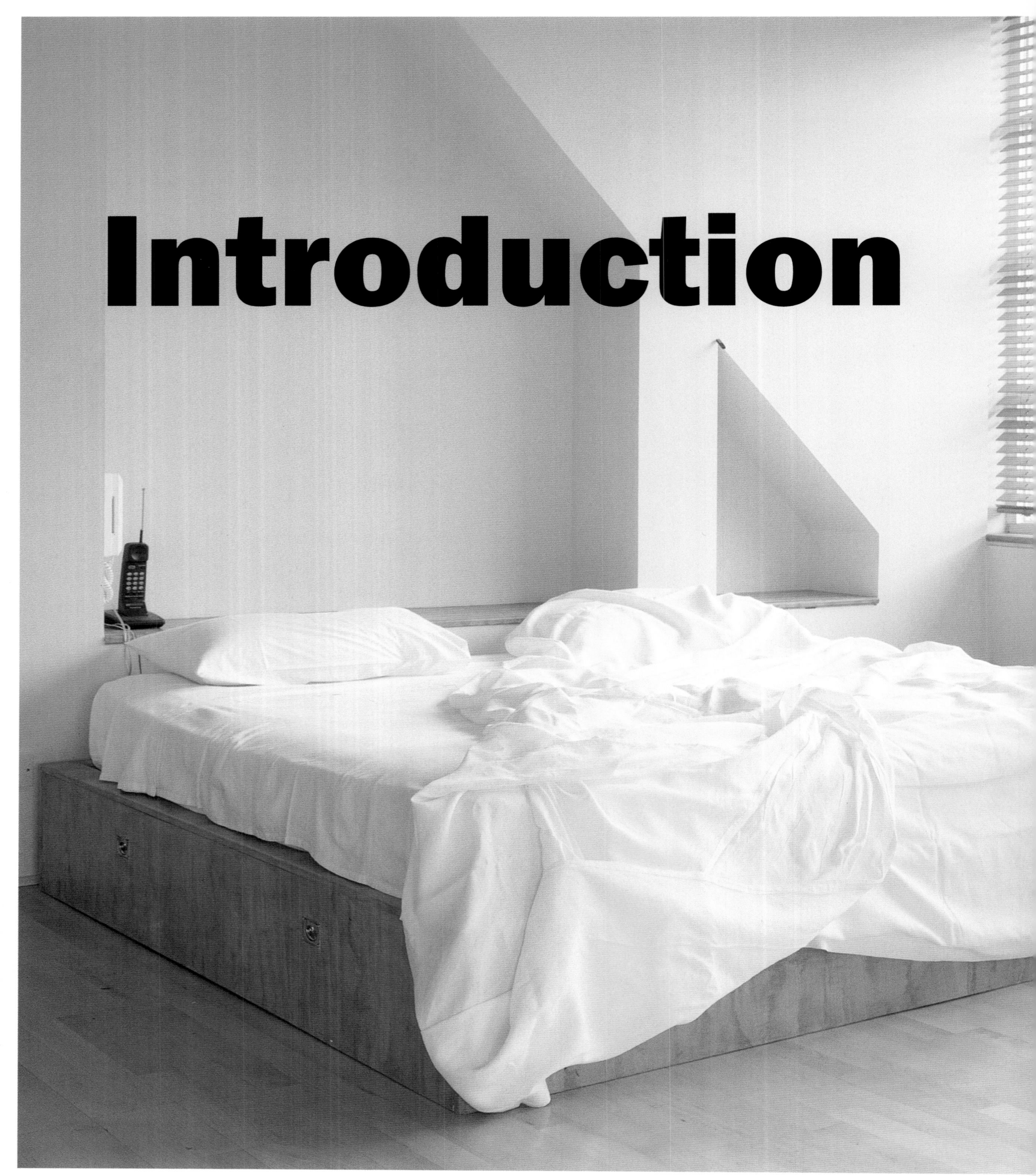

The floor is a key element of the interior, literally the basis of every room. After walls and ceilings, floors represent the largest surface area in the home, which means that new flooring can entail a substantial investment of time, money and effort, so it is important to get it right.

Equally important, the floor is a surface with which we are in more or less constant physical contact. Depending on its context, a floor may need to be comfortable, warm, quiet, safe, durable, easy to maintain or meet any of a number of other physical requirements. At the same time, what the floor looks like – its basic character, colour, pattern or texture – will inevitably set the tone for a decorative scheme. Technically, floors may be part of the background, but very few other elements of the interior have the potential to create such an impact on the way we live.

While it is easy to stand back and admire fresh paintwork or a new sofa, what goes on underfoot often escapes our attention. Yet a beautiful floor can do more for a room than almost any other aspect of decoration or furnishing. If you take the time and trouble to get the flooring right in your home, the rest will fall into place much more readily.

Choosing the right flooring is a decision which cannot be taken lightly — by and large, mistakes cannot be simply glossed over or hidden away. Before you begin to research alternatives, consider how all the areas in your home relate to one another. Ideally, flooring should be planned and chosen for the entire home rather than for one particular area at a time. If that is impossible, it is still important to choose flooring for one room with reference to what already exists or what is planned in other areas. This is not merely to ensure that there are no awkward clashes where one floor meets the next, but that the entire experience of moving through from one area to another is a natural and enjoyable progression.

Top: Warm terracotta tiles have a rustic charm. Handmade versions are full of character and age attractively.

Above: Tiling offers the opportunity to create eye-catching floor-level patterns such as this radial design.

Left: Terrazzo is a striking material, ideal for interiors in warm climates. It may be laid in situ or in the form of tiles, as here.

Far left, above: Wooden flooring has an inherent rhythm, both in the pattern of the graining and in the length and width of individual boards.

Far left, below: Blond hardwood is a classic contemporary floor, combined here with glossy black resin.

Far left, centre: Concrete need not be brutal. This matt-textured concrete floor has the look of brushed suede.

Material quality

We tend to think of colour, pattern and texture as separate elements of design. To some extent this makes sense, particularly in the context of decoration, for example, where one aspect may well dominate the others. A red wall, for instance, may look subtly different with a matt or sheen paint finish, but there is no disguising the fact that the colour is the key issue.

That distinction of separate design elements cannot apply to floors. There are, of course, certain situations where a splash of colour or a bold pattern is exactly what is called for. But choice of flooring chiefly concerns materials and their intrinsic qualities. The colour of a hardwood floor may be one of its attractions, but most people do not choose a wooden floor on the basis of its colour alone. They make their selection after considering a number of related characteristics: the texture of the grain, the pattern of the boarding or the way the floor feels and sounds underfoot. Success with flooring often depends on an awareness of the way in which the elements work together to create a sense of character and style.

There is no escaping the fact that many of the flooring materials we find most beautiful and evocative do not come cheap, frequently because they derive from natural, often limited, sources. Natural materials add a fourth dimension to colour, pattern and texture — that of time. By and large, natural materials wear and age well, and many look better and better as the years go by. The same cannot be said of most synthetics; either the material is relatively impervious to change, which gives it a lifeless, static quality, or it degrades and looks tawdry as it wears.

This is not to argue against the use of artificial flooring in every circumstance. Synthetics have persuasive practical advantages, the most eloquent being economy. Many artificial materials have been deliberately designed to give the appearance of their more expensive natural counterparts, but they cost far less. Labour-saving is another important criterion, and many types of synthetic flooring are simpler to install and easier to maintain. On stylistic grounds alone, however, synthetic flooring generally works best where it is a positive choice, rather than an obvious stand-in for something better.

Basically, it is a question of integrity. There is a vast difference between sheet vinyl which has been patterned and embossed to resemble brick and the same material in a simple chequerboard design. The latter may well suggest the effect of a black-and-white marble floor, but it is not setting out to deceive you.

In this context, lateral thinking can take you further than simply substituting a down-market simulation for the real thing. For many high-price solutions, there are more affordable alternatives which

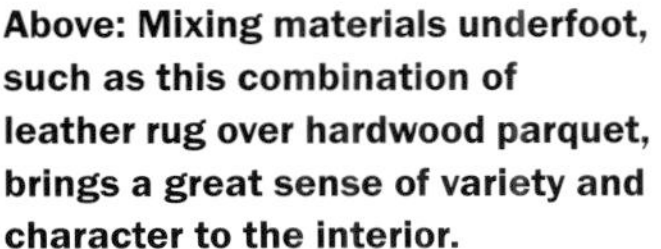

Above: Mixing materials underfoot, such as this combination of leather rug over hardwood parquet, brings a great sense of variety and character to the interior.

Above, right: Wide panels or planks have an expansive look, used to great effect in large spaces. This spruce floor has been simply sanded and sealed to reveal the grain of the wood.

Right: New, and sometimes surprising, flooring materials are emerging all the time. This graphic-looking 'area rug' is made from woven paper twine.

Left: Limestone is a supremely elegant material that suits contemporary and period interiors alike. In cold climates its chilliness can be mitigated by underfloor heating.

can work surprisingly well. So if your heart is set on marble for the hall, but your bank balance won't bear the burden, try to identify which of the particular qualities of marble you find most attractive. If it is the smoothness and coolness of the stone, ceramic tile may provide an evocative alternative. If it is the association with period grandeur, marbleized floorboards might be the answer. If it is the sheer luxury of the material that appeals, you could either run a narrow border of marble tiles around a quarry-tiled floor or have them inset as threshold strips to give an elegant lift to more everyday surroundings.

When selecting flooring materials, do bear in mind that patterns, colours and textures can look very different under different lighting conditions and at different scales. Tiny samples or catalogue pictures can be very misleading — as can showroom lighting.

Architectural character

Different materials also arouse specific responses as a result of their history of use in the interior. Subconsciously or not, this heritage aspect often influences our choices. Over the centuries, floors and floor coverings have been made from a vast range of materials, from mosaic to matting, from painted oilcloth to rammed earth. Ideas which seem fresh and original today often turn out to have quite a respectable pedigree. Natural fibre coverings, for example, the style insider's choice since the mid-80s, have their origins in the layers of rush that were strewn on the floors of English medieval hall houses, while George IV appreciated the effect of richly decorated staterooms complemented by the simplicity of coir matting as long ago as the turn of the 19th century. And mosaic has been used since the time of the ancients.

The Renaissance marked an appreciable shift in attitudes to the interior and its design. Classical systems of design and decoration, which had originated in the ancient civilizations of Greece and Rome, were rediscovered and adopted wholesale by architects, artists and craftsmen, first of all in Italy. This influence gradually spread, reaching the northern parts of Europe by the 18th century and then the New World too.

A central tenet of the classical style was the notion of conceiving architectural space, its decoration and furnishing, as a single unified scheme of design, rather than a series of distinct elements. In the homes and palaces of the very wealthy, imported marble and other decorative stone began to make an appearance as flooring materials, often fashioned into exquisite patterns, while in the

Nevertheless, in most households up until the beginning of industrialization floors tended to be fairly basic and rudimentary. Buildings were largely fashioned from whatever materials were locally available – from stone quarried in the region, and oak or other timber from nearby forests. With very few exceptions, the floor was principally a structural element, nothing more, and carpets, which were rare and expensive items brought back from the Orient or Near East by traders, were prized treasures to be hung on walls or draped over tables.

grander houses timber flooring often took the form of geometric inlay known as marquetry or parquet. The floor began to be considered and designed in sympathy with other architectural features – mirroring the intricacies of a classical ceiling, for example, was one device that became the hallmark of the English architect Robert Adam.

With the establishment of carpet-making at Wilton and Axminster in south-west England and the court manufactures in France, carpet became more readily available, and began to be used on the floor rather than displayed as a wall hanging. But for the majority, however, carpeted floors remained a luxury. In Britain, the elegance and refinement of the Georgian interior were more commonly complemented by relatively unfinished timber floors laid in boards, cleaned with sand, the light tone of the wood marrying well with the cool pale colours which are characteristic of 18th-century decoration. The same type of simplicity is evident in classical Scandinavian interiors of the period, and in American Colonial and Federal houses, where the boards tended to be wider and were sometimes painted or decorated with simple motifs or stencilled patterns. Rag, hooked or braided rugs, flat-weave 'ingrain' carpet and painted floorcloths were more usual coverings than fine carpet.

Far left: Pale hardwood strip marries well with the light airy structure of a conservatory.

Bottom left: A pristine oak floor gives a lift to period panelling.

Centre left: Rubber comes in a wide range of vivid colours.

Centre right: A modern art rug provides a softening layer in a minimal interior.

Left: Decking and hardwood.

Below: Cool, stylish terrazzo is ideal for hot climates.

With the progress of industrialization, however, carpet became more affordable and widely available. New synthetic pigments and dyes were developed and in Britain the Victorian love of ornament and pattern was displayed in vividly coloured and richly detailed carpets. These were often laid wall to wall, or with only a small margin of exposed wooden floor, usually stained a dark colour to tone with the rest of the woodwork. Alternatively, the margin might be filled by a utilitarian covering such as drugget, oilcloth or linen. A woolly rug or animal skin took pride of place before the hearth.

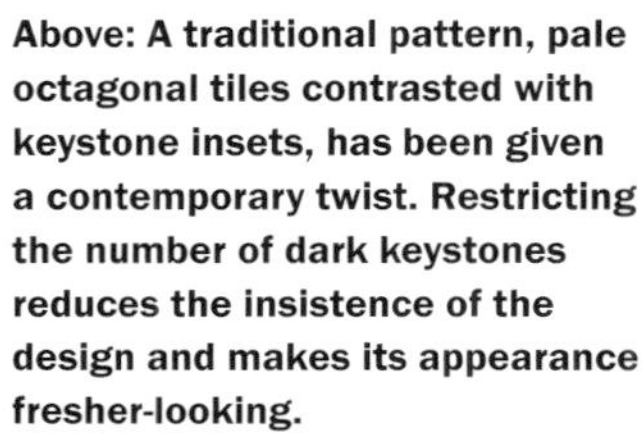

Above: A traditional pattern, pale octagonal tiles contrasted with keystone insets, has been given a contemporary twist. Restricting the number of dark keystones reduces the insistence of the design and makes its appearance fresher-looking.

Right: Simple, integral flooring, such as this hardwood strip, keeps the attention focused on the organic shapes and bright colours of the furniture. The reflective sheen of the flooring maximizes the effect of natural light spilling through the venetian blinds. A softer, more light-absorbent flooring, such as wall-to-wall carpeting, would have produced a much less lively and spacious feeling in the room.

On ground floors, in halls and in service rooms, where floors had always been made of heavy, durable materials such as fired earth, stone and brick, new materials began to be used. Encaustic tiles, with their Gothic motifs, expressed the 19th-century fascination with medieval art and were a popular treatment for hallways and porches in Britain from the mid-century onwards. Oilcloth was a similarly serviceable and cheap alternative, often exuberantly decorated in simulation of other more expensive materials. Its successor, linoleum, appeared around the turn of the century.

The introduction of linoleum coincided with a new concern to create more hygienic surroundings, where dust and vermin could not lurk in heavy drapery or carpeted floors. This desire for a clean sweep meant that wall-to-wall carpeting was replaced by smaller rugs laid over polished or stained boards which were easier to maintain. Gradually, as the century turned, the trend was towards more lightness in the interior: parquet and wood-block flooring became fashionable; and, particularly under the influence of the Arts and Crafts movement, quarry tiles and brick became more commonplace.

Throughout the 20th century the range of synthetic floors has increased at a great rate, while natural flooring materials from all over the world have also become more widely available. Simultaneously, a wealth of stylistic influences has come into play — influences as diverse as the Mediterranean villa and the traditional Japanese house, the Cornish cottage and the Manhattan loft and even, in the industrial aesthetic of high-tech, the factory or commercial office. It is perfectly possible to create any style that takes your fancy. Given this breadth of choice, it can often be difficult to know where to begin.

Below: Concrete has emerged as a stylish contemporary flooring material, finally shrugging off its image of ugly, grey utility. These concrete tiles have been varnished to give them a mellow earthy colour reminiscent of quarry tiles.

Below right: Cool limestone makes a sympathetic choice for a modern bathroom. The material is repeated in the bathtub itself and a neat stone skirting provides an elegant finishing touch. If you use stone in areas that are likely to get wet, choose a slightly textured finish to prevent floors from becoming excessively slippery.

Bottom right: Original period floors are considerable architectural features, well worth preserving and restoring. This time-honoured pattern is a classic design that combines pale stone octagons with small insets of black stone. The typical combination is pale or white marble with black slate.

One useful starting point when it comes to creating a sense of style is to consider the period and architectural character of your home. Of course there is no law which says that you must preserve original features and forms of decoration right down to the last doorknob or fingerplate, but a basic sympathy for context can help guide your choices. And although a floor covering is a renewable, superficial treatment, the floor itself is literally built into the fabric and structure of the house, so respecting or restoring what is already there is often the most successful approach to adopt, if your home has any degree of architectural distinction worth preserving.

The growing interest in period styles and historic interiors means that it is now easier than ever before to find the information you need and to source the materials necessary for restoring or replacing an original floor. While not everyone is going to discover Victorian encaustic tiles hiding their glory under the cracked linoleum in the hall, it is far from uncommon to find relatively intact original floors buried under more utilitarian coverings. Brought to light, refinished and with any missing elements replaced, such floors contribute a unique and satisfying sense of character that more than offsets the time and effort spent on their restoration.

But you don't have to live in the country to relish rugged materials and natural colours or nubbly textures – and neither do you need to own a fine period property to enjoy old stone flags or hardwood parquet. If you have an overall sense of your own tastes and what you are trying to achieve, the result will inevitably have greater vitality than following stylistic blueprints to the letter.

Scale and proportion

Most of us would like to live in more spacious surroundings than those we already inhabit. But you don't have to physically enlarge your home to increase the feeling of space within it. The right choice of flooring is one important way in which you can create a sense of openness and expansion.

In very small flats or houses, the most straightforward approach may be to run the same flooring more or less throughout. This has a unifying effect where space is cramped and provides a basic simplicity which helps to counteract lack of scale. In small homes, a significantly greater proportion of space is often taken up by circulation areas such as hallways, landings and stairs, which means that you are more aware of the transitions from room to room and area to area. Running the same flooring throughout helps to lessen the impact of these transitions simply because you draw less attention to them.

The obvious risk with such an approach is blandness. But whereas yards of neutral-toned carpet would undoubtedly be boring and uninspiring, there are many more livelier alternatives that can still enhance spatial quality without looking dull. The secret is to opt for flooring materials which have their own innate sense of liveliness or interest. In such situations, natural fibre coverings, such as sisal, jute or coir, are much more appealing than cut pile carpet because the nubbliness of the weave and texture of the fibres provides an added dimension.

It is rarely possible to extend the same flooring into absolutely every corner of the home. Hard-working areas such as kitchens and bathrooms tend to require more practical floors than those designed for general living spaces. Nevertheless, the quality of space will not be undermined if you keep to the same tonal range when flooring these areas. Linoleum, vinyl and ceramic tile now come in such a wide variety of colours that making a good match with the main flooring is not difficult.

In open-plan spaces or multi-purpose rooms, or where there is more than one level, diversity rather than unity may be what is required. Here, the choice of flooring can help to define different areas of use and break up the monotony of a large surface.

Practicalities aside, large open areas can look rather amorphous without some sense of definition to humanize the scale. To take an extreme example, a loft or converted warehouse is potentially among the most exciting of all living spaces. But for the space to be exhilarating rather than intimidating, it has to be treated as a series of related areas rather than one vast stage set with furniture and fittings marooned in isolated groups. So where you want to retain the sense of openness without inducing a feeling of

Left: Textural contrast makes an important contribution in interiors where backgrounds must remain in the background. Here loose-laid sisal matting, with its coarse, nubbly weave, is a perfect complement to the bleached, distressed surface of reclaimed oak boards. Where flooring materials are used in combination, detailing is crucial. The neat jute edge binding on the sisal matting both prevents fraying and makes the effect look well-considered.

Below: In minimal interiors, simplicity can be deceiving, demanding a high degree of perfection in execution and finish. Where the wide planks of hardwood flooring meet the plane of the wall, the wall is carried over on an inset beading instead of being finished with more traditional skirting. The effect is to 'float' the walls over the floor and accentuate the pure sculptural quality of the architecture.

agoraphobia, using different flooring materials in different areas introduces a suggestion of enclosure that will make the space easier to live in.

Other factors have an impact on our perceptions of scale and proportion. As in all aspects of decoration, pale or neutral tones increase the impression of light and space, whereas darker shades provide a greater sense of warmth and enclosure. In a similar way, the reflective sheen of harder materials, e.g. polished wood or tile, can enhance a feeling of spaciousness, while matt surfaces, such as pile carpet, which are light absorbent, have the opposite effect.

The way you lay flooring which is composed of repetitive elements, such as boarding or tile, also has a bearing. Strips of hardwood laid parallel to the length of a room lead the eye onwards. Herringbone patterns made with brick, wood block or parquet are more dynamic than basketweave designs, because they introduce a sense of direction and forward movement. Using tiles of a scale to suit the space in which they are laid is important. A widely spaced grid of big tiles or blocks looks expansive in a large area; in a small room, large tiling only calls attention to the lack of floor area.

But all rules are made to be broken. In a small confined space, the answer may be to acknowledge the limitation of scale and play up to it, with jewel-bright colour or geometric patterning. Strong colour is uplifting, but you can have too much of a good thing and what works in relatively small doses may well be exhausting or simply too dominant in a large measure.

Top left: Cool limestone has a monumental quality paired with luminous blue wall tiles.

Top right: A striking floor is composed of narrow oblong blocks in a mixture of different woods.

Above left: Pale French limestone flooring is extended from the main living space right out into the garden terrace to dissolve the boundaries between indoors and out. In situations where you cannot extend precisely the same flooring material into outdoor areas, an equivalent effect can be achieved either by choosing a more durable material from the same family, such as stone or wood, or by choosing a material of a similar colour or texture.

Left: A dynamic, directional effect has been achieved by laying small square wood blocks on the diagonal, leading the eye through the space. This effect is further enhanced by the stripy patterning of the warm rosewood veneer.

Above: In open-plan spaces, the challenge is to retain the feeling of spaciousness while demarcating distinct areas, a task achieved here by flooring the kitchen area in rubber to contrast with hardwood elsewhere.

schiele
paolo uccello
courbet
durer
GIORGIONE
MANTEGNA
GIOVANNI BELLINI
MEMLING
pisanello
Bacon · Freud
VERMEER
DES ANGES
La maison de Freud
Au mépris des règles

Connections and combinations

Outdoors, we tend to be acutely aware of changes in terrain. Part of the pleasure of a country walk is the way different textures and surfaces are encountered underfoot. The scrunch of gravel, the springiness of turf, the soft rustle of leaves in the woods all contribute to the experience. Conversely, we tire quickly of walking in urban areas if the surface is unremitting tarmac or concrete.

The same basic enjoyment of surface can be provided when you make your choice of flooring for the home. Sometimes, especially where space is very cramped, it is necessary to keep to the same flooring throughout the home but in most cases, a combination of different materials helps keep the space alive. Varying the type of floor from area to area does not imply that you have to assault the eye with clashing colours and patterns or make abrupt changes of style. It simply means being aware of the way in which changes of surface or texture can add vitality and interest, a subtle shift in gear from one room to the next.

In an open-plan area, one of the simplest and most straightforward ways to achieve the required distinction is with a rug or series of rugs. In a large room, a rug will help to draw a seating arrangement together; in a combined living/dining space, a rug under the table and chairs will add a touch of formality and graciousness. It is important to choose the right size of rug for the scale of the room. Small rugs scattered about in a big space will only look lost.

A further dimension can be achieved by varying the flooring material. There are often good practical reasons for combining different types of floor in the same space. In a kitchen/dining room or where a kitchen is included within the main living area, a robust and easily maintained floor is usually the best option for the hardworking part of the room – the cooking and food-preparation area – while a softer floor may be preferred in the rest of the space.

With this approach, it makes sense to find a natural break. The dividing line can be determined by the architecture or structure of the room or by the way the space is used. Where two rooms have been knocked through, the position of the former partition wall is a good place to change from one type of floor to another, particularly if there is an archway or a margin of retaining wall on either side of the opening. If a kitchen has been built into one end of a living room, behind a counter or half-height divider, the most obvious solution would be to run the main flooring up to the counter and treat the kitchen floor differently.

The most discreet way of combining floors in open-plan areas is to vary the material without varying the tone. A light hardwood floor in a dining area would make a seamless partnership with light ceramic tile in the kitchen end of the room, for example.

Top: Material mix can be as simple and straightforward as a scatter rug over reclaimed floorboards.

Above: Echoing the combination of materials in kitchen units and shelving, metal kitchen flooring meets wooden floorboards.

Left: Subtle variety of textures underfoot: wood meets concrete inset with a generously scaled coir mat.

Top right: Varying the material without varying the basic colour prevents uncomfortable-looking contrasts. Here stainless-steel sheet is combined with silicaceous pebbles set in resin, providing an unusual textural contrast.

Above right: Slate laid in staggered rows gives onto the airy expanse of maple floorboards.

Right: Strong contrasts can be equally as effective as sympathetic matches. Here brick paviors are crisply offset by a border of white marble, an economical way of using an expensive material.

But it can be equally effective to make a bolder contrast of material, colour or pattern. Where there is no natural break, or the way you use the space does not conform to a structural division, you can use flooring in a freer way to provide a sense of definition. This often works best where you retain the same flooring material throughout, but vary the colour or pattern. Sheet lino is a good material for this type of treatment: you can use curving contours of contrasting colour to signal the change from one area to the next. A cheap and effective way of defining areas is to paint or stencil a border on floorboards to create a space within a space, or introduce a patterned tiled floor within a greater expanse of plain tiling. On the whole, however, bold contrasts do not work well with carpet. It is nearly always better to avoid any situation where carpets of differing colours, textures or patterns meet without some intermediary shift to a different material.

Changing levels also provide a good excuse to vary floor treatments; in fact, a change of level that is not marked out in this way can look distinctly awkward in certain circumstances. Then there is the safety aspect to consider. Where part of a room is a

Above: A study in elegant contrast is provided by this unusual flooring combination: an inset panel of recycled 'ironbark' timber (an Australian hardwood) is set within saturnia stone, which is a type of unfilled travertine.

Above centre: A deep-pile rug in grass-green makes a contrast of colour and texture with the old polished floorboards. Wooden floors can be noisy; rugs not only add comfort but also deaden the sound of passing feet.

Above right: Slate comes in evocative moody shades. These dark slate tiles make a striking contemporary floor and a graphic foil for the crisp white walls.

Right: Tonally similar, sisal in the hallway meets herringbone parquet. Metal edging strips give a neat finish where the two materials adjoin.

Far right: An elegant study in neutrals combines a plain wool wall-to-wall carpet with a rug only a fraction lighter in tone.

Left: A woven stair runner reduces slipperiness underfoot. Stair carpets should always be secure.

Right: Changes of level are an opportunity to change material. Here a metal staircase descends to a plinth of stone, which steps down to the main beechwood floor.

Centre right: A quarry-tiled kitchen floor is a sympathetic match with hardwood and a flatweave runner.

Far right: Textural contrast is given by a bubbleweave rug over slate.

Above: Nothing is more theatrical than glass flooring. These superbly detailed glass stairs have an unbeatable sense of drama.

Above: Stone can be chilly underfoot. This tufted linen rug laid under the bed over a limestone floor provides additional comfort for bare feet.

step up or down from the remainder of the space, you can emphasize the transition with a change of flooring material, and reduce the risk of anyone missing their step.

Stairways indicate another change of pace and offer a logical place to switch from one type of material to another. The theatrical sweep of a staircase descending into a hall can be emphasized by just this sort of contrast but even a modest flight of stairs looks well in a different finish.

Another trick is to combine materials over the same floor area. This is an excellent way to create textural variety and it can also provide an economical yet effective means of using a more expensive material in smaller quantities. There are certain classic, complementary combinations, from the traditional white marble with inset diamonds of black slate to the more homely partnerships of brick or terracotta and wood, but whatever materials you put together should show some basic affinity, either in terms of style or character.

Floor-level detail

Most floors occupy a relatively large surface area, but that does not mean you can afford to overlook the details. No matter how beautiful a floor may be, the full effect can be severely undermined by paying insufficient attention to what at first may appear relatively insignificant elements – such as junctions or edges, which should always look considered and neat. Finished edges not only look better, they perform better, too. Unprotected edges are more likely to fray and degrade and may even trip you up. Certain materials can be successfully butted up against each other, provided they are of similar thickness, but it is usual to cover the join with a strip. Aluminium or brass cover strips are the standard means of finishing the edge of carpeting, particularly where it meets a new material. If you are running a carpet through several rooms, however, a seamed junction in the doorways would be less obtrusive. Bevelled wooden strips can be run at the edge of hardwood floors or in thresholds. For a more considered and elegant approach, you can fashion threshold strips from a contrasting material, say a wide strip of wood or marble tile.

Above: Luminous blue and white glazed tiles provide an eye-catching decorative detail inset between terracotta paving.

Above: Floor-level lights set in limestone follow the contour of a curved partition wall for night-time drama.

The perimeter of the floor also merits attention. The flooring should be neatly laid into the angle where it meets the wall to give a clean defining line. This may entail removing the skirting boards during installation. If your skirting boards look battered and chipped, strip and refinish them. (Devotees of minimalism like to do without such mouldings altogether, stopping the wall just short of the bottom on an inset bead, so the wall appears almost to hover over the surface of the floor. However, a skirting board does protect the lower portion, so you need to consider their practical advantages.) Floor coverings should also be laid right into cupboards and closets. You may not notice the difference most of the time, but whenever the doors are opened, the lack of finish will be all too apparent.

Above: Concrete in the kitchen area meets hardwood parquet in the dining area, a deliberate shift from functional to aesthetic.

Centre: A glossy expanse of black resin flooring makes a striking contrast to oak boards.

Left: Glass flooring lends translucency and a sense of glamour to the interior. Glass must be individually specified.

An attractive floor invites floor-level living. The obvious point of reference is the traditional Japanese house, where tatami mats provide the springy cushioned base for sitting, sleeping and eating. In the West, we tend not to spend quite so much of our lives on the floor, but as our homes and lifestyles have become less formal and constricting, the focus has naturally shifted, literally towards the floor.

Low-level furniture, such as futons and divan beds, low coffee tables and floor cushions encourage the use of the floor as a place for relaxing. Decorative floor-level displays also spell out the same message. Collections of beachstones, groups of indoor plants, and large urns, bowls or platters can all provide a dramatic focal point at floor level, particularly when picked out by a concealed uplight or free-standing lamp. Natural sites include around and in front of a hearth, or in corners out of the way of mainstream traffic.

Left: Wooden floors of all descriptions are extremely versatile, both stylistically and practically, as this maple flooring demonstrates.

Right: Original hardwood parquet is a floor well worth preserving and restoring. Being solid, the wood can be stripped and refinished.

Below right: Water-resistant and easy to maintain, tiles make ideal flooring for bathrooms and kitchens.

Bottom right: A pale French limestone makes a cool and sophisticated kitchen floor. It is important to make sure that the subfloor can bear the weight of the stone. Some stone can be supplied in lighter tile form.

Far right: Textured non-slip vinyl in eye-catching pink provides a safe covering for a metal stairway.

Assessing your needs

Style, character and proportion; colour, pattern and texture – the aesthetics of choice – are crucial factors in selecting flooring. But the physical attributes of different materials and treatments must also be considered to ensure the floor performs as well as it looks.

In the following chapters, the properties of each type of flooring are covered in detail. Before you weigh up alternatives, however, it is worth reviewing some of the broad practical issues involved, then assessing your own particular requirements. Those needs will vary from household to household, and from room to room.

Safety Many avoidable accidents happen in the home. Of these, falls and tumbles can be among the most serious in their consequences, especially for young children, the elderly or the infirm. The hazards may derive from the nature of the flooring itself, from where it has been laid, or from the way it is treated and maintained.

Inherently slippery surfaces, such as terrazzo, polished linoleum, waxed wooden floors, clay or glazed ceramic tiles, present an obvious danger in certain locations. In bathrooms and kitchens spills of water or grease can add to the risk of taking a tumble: carrying a hot casserole from the oven or stepping out of a bathtub are the type of manoeuvres which demand sure-footedness. And if you like to go about the house in stockinged feet, the risk of slipping increases.

Changes of level increase the chances of missing your step, so special care must be taken to choose the right materials and coverings for stairs and landings. Slippery flooring is not recommended for stairs without the use of non-slip insets or nosings and some materials, such as very coarse natural fibre coverings or shag-pile carpet, are best avoided altogether. Handrails should be provided for the elderly and any stair carpet or cover secured firmly in place to reduce the risk of tripping.

Poor maintenance can be a cause of trouble. Loose floorboards should be nailed down and wobbly tiles securely adhered. There is no alternative but to replace frayed stair carpets to prevent heels catching in the worn places.

Highly waxed or over-polished floors are dangerous and it is well worth living with a less glossy finish for the sake of safety. Also, unsecured rugs are a hazard. All rugs should be laid over non-slip matting to prevent their transformation into flying carpets. This applies equally to the rag runner by the bedside as to the Oriental in the living room.

Maintenance Obviously, all floors require regular cleaning to avoid a build up of dirt, grease and grit; some demand additional treatment in the form of intermittent polishing, resealing or waxing. Some types of material stain readily and irrevocably; others are more resistant. Your choice of flooring should be influenced to some extent by how much effort and expense you are prepared to invest in its care, as well as how appropriate the material is to the intended location in the home. Different materials need to be cleaned in different ways and it is always advisable to use products and techniques recommended by the suppliers or manufacturers.

Left: The colours of these Provençal-style tiles have been picked up in the soft furnishings.

Below left: Mosaic tiling makes a practical entranceway.

Below: Traditional stone flooring in a classic pattern has a period flavour in a formal dining room.

Bottom: A raised hardwood floor accentuates the airy qualities of an open-plan loft.

Centre: A felted wool rug laid over stripped wooden boards adds subtle textural contrast.

Far right: Tiny pebbles embedded in resin make an unusual, hard-wearing textured surface.

Wear The degree to which a type of floor will wear is a function not only of the traffic it receives, but also of how fundamentally robust it is. The concept of 'wear' is relative. Wood, terracotta and other natural materials wear well, acquiring an attractive patina of use. Subject certain synthetic flooring to the same treatment and the result may well be an eyesore. Perception of wear can also be affected by the colour, pattern and texture of the material. Obviously, a pure white wool carpet is going to show every mark, whereas a flecked berber would be more forgiving.

Traffic is heaviest in circulation areas in the home – entrances, hallways, stairs and landings – and lightest in bedrooms. In addition, there will be variations in traffic within any one space.

For example, the area beneath the dining table, where chairs are pushed back and forth, receives much heavier wear than the perimeter areas of the room.

Certain precautions will mitigate the effect of undue wear. Few types of flooring are proof against the intense pressure points of stiletto heels or the castors or tapering legs of heavy furniture. Protective cups will help to spread the point load of furniture legs – but the only remedy for spike heels is a change of footwear. Extra protection will also be needed in entrances and other places where gritt can be tracked indoors. A wide coir mat inset across the full breadth of the hallway will catch a good deal of loose debris and absorb damp, both of which break down seals and finishes.

Stair carpets fixed by rods are more practical than fitted stair carpeting, since they can be lifted and turned to spread wear evenly. Any areas which are likely to see a high degree of wear should ideally be floored with a material which is easy to renew.

Comfort Floors which have a degree of resilience or 'give' are less tiring on the legs and feet and therefore more comfortable than those with little resilience. Wood, cork, cushioned vinyl, linoleum and low-pile carpet are all fairly springy underfoot, the more so if they are not laid over ungiving screeded or concrete subfloors. Hard materials, such as stone, brick and clay tiles, have next to no resilience and are both less comfortable to stand on for long periods of time and more likely to cause breakages if anything is dropped on them.

It is worth giving some thought to texture, too. Soft or smooth materials are kinder to bare feet and young knees than scratchy, rough or embossed surfaces.

Noise and temperature control Soft floors are the quietest way of muffling sound within a room and reducing the amount of sound that travels between floors. Hard floors amplify sound, which may cause particular problems at upper levels. A layer of sound-proofing can be incorporated under the main flooring to reduce noise further.

Continuous flooring, such as sheet linoleum or carpeting, insulates against heat loss by excluding draughts. Soft materials, such as cork and carpet, also warm up more quickly. Stone, terrazzo, clay, concrete and brick, on the other hand, are all chilly surfaces, which can be a positive advantage in a hot climate or in areas where a cooler temperature is desirable, such as in a larder or pantry or a kitchen.

Cost Price remains one of the most critical factors in the selection of flooring, but a simple cost comparison per metre can be misleading. Make sure you calculate the true finished cost of the floor you want to lay. To the price of the basic material you must add the cost of installation, plus any sub-floor preparation or materials such as underlay and junction strips, the cost of sealing or finishing the floor surface, and its aftercare or maintenance (see pages 180–81). In general, be realistic: if you find you cannot afford the type of floor you really want, it is best to rethink the whole scheme, rather than settle for a poorer quality version of your first choice.

Another factor to consider is the potential for refinishing. There is often little to be done about a worn synthetic covering. A worn wooden floor, on the other hand, which may well have been more expensive at the outset, can always be resanded and restored, prolonging its life and your investment. Similarly, the look of brick and terracotta tiled floors positively improves with age.

Most cheaper types of flooring have correspondingly shorter lives, so consider how long you need the floor to last. It is also worth thinking about how long you expect to stay in your present home; if you are renting, or there is the possibility of moving in the near future, area rugs, carpets, loose-laid matting and other types of flooring which can be taken up when you go may be the best way to spend your money.

Opting for the cheapest flooring, only to find you have to replace it sooner rather than later, is a false economy. Others include doing without proper professional help where it is required and skimping on preparation or underlay. There is little point in covering up a worn, pitted or uneven floor studded with nailheads and hoping for the best. The poor quality of the underlying surface will quickly show through and cause wear and tear on whatever material you have covered it with.

Hard floors

Mellow brick or opulent marble, utilitarian concrete or glittering mosaic, hard flooring materials offer an immense breadth of stylistic choice to suit any scheme. Colours span the spectrum from natural neutrals to intense glazed hues; textures range from silky smooth to rippled and riven; and patterning encompasses everything from speckles, flecks and veining to representational motifs.

Yet all hard floors share certain common characteristics, many of which at first sight may seem positively disadvantageous. They are mostly chilly underfoot and can amplify sound to an uncomfortable degree. They lack resilience, which means they are tiring to stand on for long periods and anything dropped is likely to break. Most hard floors are heavy and require either a solid sub-floor or additional reinforcement to bear the load; many demand professional laying. Some hard materials are brittle before they are laid, which increases the risk of damage in transit or installation; installation itself can be disruptive and time-consuming. Most prohibitive of all, many types of hard floor come with a hefty price tag.

There are, however equally strong advantages. All hard floors convey a sense of permanence and stability. The chilliness and noisiness can be mitigated by adding a layer of rugs or matting,

Above, left and far right: Hard flooring encompasses a wealth of evocative materials (clockwise from top left): sophisticated terrazzo, practical hard tiles, utilitarian concrete, homely brick, gleaming metal, cobbles and pebbles, traditional marble with slate insets, glossy resin, transparent glass, and mellow terracotta.

Right: Smooth and sleek, pale limestone tiles reflect light back into the interior and complement both period and contemporary styles of decoration. Most types of stone require sealing to prevent staining and discoloration.

although in hot climates or the heat of summer coolness underfoot can be positively welcome. A stone or brick floor also helps provide natural refrigeration in larders and store-rooms.

The bulk, weight and attendant difficulties associated with the installation of hard floors have increasingly been addressed by modern suppliers and manufacturers. Many types of stone, for example, are now available in the form of tiles, which are far easier to handle than slabs, and in thinner sections which create less load.

Hard floors are built to last. Properly maintained, they will take a lot of punishment, which makes them ideal for areas of heavy traffic, such as hallways and passages, and those which connect directly with the outdoors. They may cost more in terms of time, money and effort to install, but once laid, they have the potential to last forever.

But the most persuasive argument of all remains the aesthetic one. Stone hewn from the earth has a character like no other material; fired earth is warm, rustic and homely; ceramics alive with luminous colour bring a sense of uplift to any surroundings. Even concrete or metal can be elevated from utilitarian status into a work of art.

Brick

Brick is the oldest manmade building material, and as reassuring as the earth from which it derives. It is both natural and regular, standardized into comfortable units scaled to the hand, and innately domestic.

In the West, brick began to gain in popularity as a building material as the need lessened for highly fortified castles built of stone, and instead smaller, more welcoming manor houses began to appear. In early English manors, brick was often directly substituted for stone: in Norfolk, Oxburgh Hall, which dates from the 15th century, is entirely built of brick – walls, floors, stairs and even a shaped brick handrail. Cheaper and easier to use than stone, brick also proved a safer material than timber. In London, after the Great Fire of 1666 graphically demonstrated the inherent risks of timber construction, brick replaced wood as the leading material for more ordinary houses and buildings.

This long history of use gives brick a familiar quality. As a flooring material, it evokes the rustic farmhouse rather than the elegant townhouse. But new brick can be surprisingly contemporary, even sophisticated. In the work of many modern designers, the partnership of brick and wood softens and domesticates clean lines and spare interiors.

Robust, serviceable and homely, brick is appealingly honest and unpretentious, inside or out. Indoors, brick is strictly a ground-floor treatment, since it needs a solid sub-floor of significant load-bearing capacity. Visually, brick works well running through a house, from hall to back door, in rooms directly connecting with a garden and in cosy country kitchens. Brick also makes a wonderful outdoor flooring for paths, steps or terraces and looks good in combination with wood and stone or softened by low-growing clumps of plants.

Above, left: In the right context, brick can look surprisingly stylish and sophisticated. Here the clean lines and regularity of the grid paving contributes strength to a contemporary living area.

Right: Old brick flooring, worn smooth by the passage of feet over generations, makes a practical surface for relaxed country living – ideal for ground-floor rooms which lead directly to the garden.

Above, right: The quirkiness of old, handmade bricks provides charm and interest, particularly in rustic settings. Second-hand ones can be sourced, but these may need to be sealed for use in areas which are subject to heavy traffic or which are likely to become wet, such as kitchens. Once a common flooring material in farmhouses and country cottages, these days weathered brick remains much in demand. Its mellow colour and homely texture lend character to traditional furnishings.

Types and characteristics

Brick is not cheap, but it is less pricey and more readily available than many other hard materials. Flooring bricks, known as 'paviors' or 'pavers', are thinner than construction bricks – they range from 19 to 51mm thick. They are also fired at very high temperatures, or 'overburnt', which makes them waterproof and virtually impervious to wear. Alternatively, you can use 'engineering bricks', which also have good wear-resistance, but are thicker than paviors. There is no law against using standard bricks as flooring, but they will eventually erode and pit into an uneven surface. Second-hand bricks have their own attractive patina of use, if you enjoy the weathered effect. For outdoor use, make sure you choose frostproof brick. Most characterful of all, and most expensive, are handmade bricks.

The versatility of brick testifies to its great practical advantages. It is a good insulator – slow to warm, but retaining heat for a long time. This particular characteristic is enhanced if brick is laid over under-floor heating. In areas which are irregularly proportioned, brick is also easier to use than stone slabs, for example, since it can be readily cut and shaped.

Flooring brick is exceptionally resistant to chemicals, impact and abrasion. It is also waterproof and non-slip (unless it is sealed or polished) which makes it suitable for hallways and kitchens.

Brick comes in a wide range of colours and surface textures. The natural warm 'earthy' buffs, browns and reds are standard, but bricks can also be blue, green, grey and speckled, as well as ridged, ribbed or rough-textured.

Laying brick indoors is a job for the professional, demanding precision and skill. Brick requires a solid sub-floor of concrete which has been cured for at least a month. It should be laid on a damp-proof membrane into a deep stiff mortar bed, and the jointing needs to be done simultaneously.

Movement can be a problem with bricks (or hard tiles) bonded to a sub-floor, due to changes in temperature or humidity. Joints should be wide enough to allow a small degree of shrinkage or expansion and special 'movement joints' may need to be incorporated at intervals throughout the floor or around its perimeter.

Patterning

The regularity and rhythm of brickwork is the source of its great appeal. There are a number of traditional patterns. Stretcher bond – staggered rows, as in walling – is one of the most common patterns. It is effective over a large area, especially where there are irregular features or obstacles. Herringbone is dynamic and leads the eye onwards. Stack bond – where bricks are laid in straight rows – is easy to lay, orderly and clean-lined. It is best in smaller, regularly proportioned areas. Basketweave, a pattern of alternating paired bricks, has a look of stability and enclosure. Coloured bricks can be laid to create a subtle chequerboard effect, or more randomly, or to create a defining border.

Maintenance

Once laid, bricks generally need little in the way of after-care, which is one of their great attractions as a floor material. Loose dust and dirt is simply swept up and the floor washed with mild detergent then rinsed. Flooring bricks or paviors can be polished, dressed with linseed oil or sealed – provided you apply sparing amounts. Such treatments, however, are by no means necessary and they increase the slipperiness of the floor. No dressings are recommended for absorbent bricks.

Above and left: Weathered red brick, its surface randomly pitted, makes an excellent hard-wearing floor, needing little more than the occasional sweeping and washing. There are a number of traditional patterns in which bricks can be laid; this is basketweave.

Right: Dark brick makes a surprisingly sympathetic floor in a period-style bathroom. It can be chilly underfoot but is not as slippery as some stones, and does not need dressing.

Stone

Stone is the quintessential natural material. Its origins date back to the very formation of the earth's crust 4,000 million years ago, and its history of use is as old as civilization itself. Stone is embedded in antiquity. Its eternal, enduring qualities convey a sense of permanence, tradition and stability.

Stone is one of the most varied of building materials in its applications. Its inherent strength and seeming immutability made it a natural choice for the finest or most sacred buildings — for shrines, temples, churches, abbeys, palaces, castles, châteaux and homes of the wealthy and powerful. But in areas where it could be readily quarried or cut, stone also defined the vernacular architecture — mountain lodges, farmhouses, cottages and barns that seem to grow out of their local landscape.

Stone floors bring an echo of this cultural history to the interior. More importantly, very few materials can equal stone for its sheer beauty. Refined, carved, polished and highly worked, or rugged and irregularly shaped, stone spans the full breadth of building styles.

The popular image of stone — cool, pale and smooth — is belied by an incredible variety of colours, textures and patterning, even within the same broad type. Limestone, for example, can range from creamy white to dappled blue, from speckly to peppered with fossils. No two slabs or tiles are alike. Even cut from the same block, each piece has its own unique surface, offering endless vitality and interest.

Like other hard materials, stone is eminently suitable for entrances and hallways, conservatories and other areas which connect directly with the exterior. Certain types also work well in kitchens and bathrooms. In living and dining areas, rectangular or square flags look grand and architectural, while irregular fieldstone has a rustic, country appeal. Of course, many types of stone are also excellent as paviors outdoors, for terraces, patios, paths and in greenhouses.

Although many devotees will swear that stone can be warm underfoot, it is by its very nature a cooler material than textile coverings or wood flooring. Its coolness can be tempered by

Above, left: A design idea used in reverse: here stone setts have been hand-painted to resemble chequered tiles, continuing the theme of black and white used throughout this bathroom.

Above, centre: Limestone comes in many colours, but the paler tones are most popular. These large slabs, laid in staggered rows, generate a feeling of light and space in a modern kitchen.

Above, right: These time-worn stone flags lend character to a hall. Stone floors are particularly suitable in any area which connects directly with the outside, as they are very hard-wearing.

Right: Timeless and classical, light-coloured stone octagons with dark cabuchon insets is one of the most familiar of flooring patterns. The style originated in stone but is much copied in other materials.

GREC

underfloor heating; once warmed, it retains heat and loses it slowly. Some types of stone are slippery, others considerably less so. All stone floors are noisy.

Solid stone ranges from heavy to very heavy, depending on its thickness. Flags and the heavier tiles require a concrete sub-floor, which tends to rule out applications at upper levels, and professional laying is recommended. Certain types, such as marble, are available in lighter, thinner and cheaper tiles, which extends the range of applications.

Stone is available from specialist suppliers and direct from quarries; some sell only flags or tiles in set sizes, other outlets will cut to order. You could also try architectural salvage yards and suppliers specializing in secondhand materials. The most expensive stone of all is antique stone, reclaimed from old country houses, farms and chateaux. With the unique patina bestowed by centuries of footsteps and scrubbing, such floors have a gentle colour and texture which is impossible to recreate. All stone is from non-renewable sources, and some types are now very rare indeed. It is costly to extract, work and transport, all of which adds to the expense: the price of a stone floor (except for tiles) generally ranges from expensive to astronomical. But for those who are captivated by stone, such limitations are borne for the pleasure of living with one of the most evocative of all materials. Stone endures, and its beauty lasts for generations.

Left and below: Hard materials, such as French limestone or flagstones make an evocative contrast with the sleek lines of a modern kitchen.

Types and characteristics

Stone is derived from three main rock types, each one defined by its method of formation. The oldest is igneous rock, created when molten rock cooled in the earth's crust millions of years ago. It is typically dense, crystalline and capable of taking a high polish. Granite is a good example. Sedimentary rock, such as limestone, sandstone and travertine, is formed from deposits of sediment and organic matter and is far softer than the igneous type. The most recent, metamorphic rock, such as marble and slate, has undergone chemical change, making it extremely hard – the result of intense heat and pressure during movement of the earth's crust.

The names of different types of stone often indicate the locality where they are commonly found. 'York stone' is a sandstone from Yorkshire; 'Bath' or 'Cotswold stone' is the name used for the characteristically honey-coloured limestone found in the Cotswolds.

Other distinctions derive from the way stone is worked. Stone is generally quarried in blocks, then cut to size and format. Faces and edges may be machine-sawn, often with a diamond cutter, or hand-worked, and finishes range from sawn and honed to highly polished. York stone and slate may be riven or split, which produces a rippled surface. As well as regular flags, oblongs and hexagons, and tiles of various dimensions and thicknesses, random natural shapes are available. Some find newly quarried stone overly reminiscent of monumental masonry, but there are pre-distressed varieties on the market with the soft, mellow look of old floors. 'Tumbling' or bushmilling blurs the hard edges of the stone and produces the random surface variations associated with years of wear.

Granite is a coarse-grained rock comprised of feldspar, quartz and mica which give it an attractive crystalline appearance. It is exceptionally hard, highly resistant to wear and chemicals, and impervious to water. Granite can be near-black, blue grey, red, pink, mottled white and its surface is typically flecked and speckly. Its natural colours may be enhanced by flame-texturing, which gives a rustic look. It is also available in a variety of other finishes, from honed to highly polished. Choose with care, though, as polished varieties may prove unacceptably slippery. Rougher textures are better for flooring; even more practical are granite setts, or paving stones which can be laid like brick. After years of use, granite can become slippery through wear and may need to be professionally refinished. It is one of the most expensive types of stone, but tiles come at a fraction of the cost and weight. Tiles as thin as 7mm can be used on bathroom floors; 10mm tiles in areas of light traffic.

Limestone is much softer than granite. Most types are fairly light in tone, ranging from warm neutral shades of oatmeal and cloudy white to dappled blue, green and grey, but there are dark and near-black varieties. Surface texture is subtle, yet lively. Faint veining, speckling, mottling or flecking is typical; occasionally fossilized traces of shells or ancient marine life are apparent. Like granite, some limestone can be 'flamed' to release the quartz content and the mottled patterning which has an antique or period look.

Limestone is found all over the world. Some of the best varieties are English and French in origin. Bath stone has been used for centuries in the finest country houses; Iscan and Blue English limestone are two rare, expensive blue varieties featured in many

Granite

Limestone

Limestone

Sandstone

historic buildings, including St Paul's Cathedral and Westminster Abbey in London. Portland stone is a creamy white version, travertine a type of very hard limestone commonly quarried in Tuscany and used for paving.

Limestone is cool and elegant. Large, even flags work in classic or contemporary surroundings, lacking the more overt stylistic overtones of marble. Some limestones wear better than others. All types are porous but may be treated to prevent staining. Limestone does not resist acid chemicals and the harder types can become slippery with wear. It doesn't come cheap: new quarried stone will cost between two and three times as much as a good mid-range carpet.

Sandstone is generally tougher, cheaper and easier to maintain than limestone. Sandstone is composed largely of fine grains of quartz, and comes in warm sandy shades, rich reddish browns, grey-greens, lavender and light pearl greys. 'York stone' is an exceptionally hard-wearing form with a riven non-slip surface. Like limestone, sandstone is porous and stains readily, but seals and polishes can make it slippery. Sandstone is available in large flags, in setts and various finishes including matt, sawn, 'tumbled' and flame-textured.

Marble is the epitome of luxury. For centuries, this cool, almost translucent stone, with its rich veining, subtle patterning and often vivid colours, has featured in the most elegant, lavishly decorated interiors. Marble was prized by the ancient Greeks and Romans as a perfect material for sculpture and building and a mark of wealth and sophistication, and much later by Renaissance architects and designers who reintroduced classical ideals to the Western world.

For as long as marble has been coveted and treasured it has been imitated. Painted faux marble floors and panelling were 18th-century substitutes for a rare and costly material; nowadays, marbleized floor tiles in vinyl or linoleum are the downmarket equivalents. Such familiarity can breed a kind of contempt, so that even authentic marble may acquire a certain tacky or vulgar look. Marble is undoubtedly grand, but in less than grand surroundings it can look curiously cheap and showy. It may convey more of the aura of a hotel foyer or corporate headquarters than refinement. More than any other type of stone, marble demands the right context. Houses which display some historical or architectural distinction are most suitable: marble is a material for making a statement. It works well in the classic graphic contrast of black and white tiling; you need to exercise greater care if you stick to a single tone – the effect can be all too reminiscent of plastic laminate.

Far left: Blue-grey granite flags make a graphic kitchen floor.

Left, centre: Granite tiles can take a high polish that is uncompromisingly modern.

Near left: Cool limestone is effectively contrasted with blue-tiled walls.

Above, left: A traditional limestone floor for a classic hall.

Above, centre: Opulent polished limestone in a minimalist setting.

Above, right: Limestone tiles make a beautiful bathroom floor. Some types wear better than others.

Below, left and right: Stone is available in a variety of colours, textures and formats, which can suit almost any location in the home – provided the sub-floor will bear the weight.

Marble

Tumbled Marble

Riven Slate

Chinese Slate

Marble is a form of limestone which has crystallized under intense pressure. It is quarried in mountain ranges worldwide, but the finest examples are generally acknowledged to originate from Italy. The purest marble is almost completely white. Nevertheless, it is often the imperfections, which are actually mineral depositis, that make the stone so attractive and appealing. These impurities result in a range of colours, from pink, red, green and brown to black, and in characteristic streaking, clouding or veining patterns. More than one colour or type of pattern may be present.

Marble is generally hard-wearing, although colours and patterns can become dulled by abrasion and traffic. The darker forms tend to be the hardest; creamy white alabaster, for example, the sculptor's marble, weathers badly out of doors. Slipperiness depends on the surface texture. For flooring, it is advisable to opt for a honed finish which obviously offers a better grip than high polish.

In addition to large sheets, marble is widely available in a variety of tile formats. Usually the tile versions comprise a thin veneer of marble backed by other materials. There are also conglomerate marble tiles, composed of chippings bonded with hard polyester resin. Sheet marble is always very expensive; tiles range from fairly expensive to more affordable.

Slate, like marble, is a metamorphic rock quarried in mountain regions all over the world, from Wales to India, from Cornwall to China. It comes in a range of beautifully dramatic colours — dark green, blue, blue-grey, red, purple and black — and tends to have a slick, wet look due to the high proportion of mica crystals layered through the stone. African slate is gloriously multi-coloured.

The presence of mica is also responsible for the fact that slate can be readily split into thin planes, a characteristic which has meant it has been widely used as a roofing material. Flooring slate is thicker than roofing sheets, but its thickness is also dependent on the way the slate is worked. Riven or split slate tends to have a slight camber, so the slabs need to be thicker and smaller than those which have been evenly sawn. Softer types of slate may have hand-worked or chipped edges, which give it a more rustic look.

Slate is by no means cheap, but it is more reasonably priced than either granite or marble and has other considerable practical advantages. Unlike marble or limestone, most types of slate are waterproof, which makes it excellent for outdoor paving as well as areas indoors which are likely to get wet. It is also very hard, wear-resistant and needs little in the way of after-care. The darker shades do, however, have a tendency to show dirt and scratches.

Right: The theatrical sweep of a stairway is enhanced by the use of slabs of dark polished slate, fanning out like a deck of cards.

Slate can be finished in a variety of textures, from riven, which is non-slip, to sawn, sanded or polished. As with other stone, the smoother finishes tend to get quite slippery when wet. In addition to regular slabs or tiles, slate is also available in randomly cut pieces which can be laid as crazy paving for a more rugged rustic look.

Pebbles and cobbles offer a simple and charming way of paving outdoor areas. In Mediterranean countries pebble floors date back thousands of years. *Krokalia* is a Greek geometric mosaic of black and white beachstones used to create courtyard pavements of immense beauty and real character. Cobbled paths, terraces or outdoor eating areas provide a lively counterpoint to planting. Increasingly the effect is found indoors too, in entrances and even in bathrooms – very massaging for bare feet, if not highly practical.

You can buy reasonably priced cobbles in garden centres or collect your own beachstones for free – although beachcombing could become a tedious affair for more than a small area. Stones should be smooth, evenly shaped and roughly similar in size. For the faint-hearted, resin-backed pebbled tiles can be bought which also work well as an a border or as contrasting insets in a flagged floor.

Reproduction stone

Purists may throw up their hands in horror, but the difficulties and cost associated with natural stone flooring have created a demand for more practical and economical simulations of the real thing. Reproduction stone has a nasty image, much of it due to the rather unconvincing 'stone effect' paviors sold in every garden centre. By and large, such artificial materials fool nobody; they simply lack any variation of pattern or texture from slab to slab and the result has a depressingly mechanical appearance. However, some reproduction stone is in a different league altogether. Faithful copies, handmade from concrete, with highly realistic colours and textures and plenty of variation from 'stone' to 'stone', lack any hint of the ersatz. The chief advantage of reproduction stone is, of course, that it is far cheaper. Its uniform thickness also means that it is much easier to lay and, sealed for indoor use, to maintain. Some flags are thin enough to lay over existing floors. Reproduction limestone, granite cobbles and York stone are among the more popular simulations.

Maintenance

Porous types of stone stain readily. Expert advice from the supplier should be sought as to whether or not sealing is advisable. Sealing tends to alter the appearance of the stone. No smooth-textured stone floor should be polished – it will become dangerously slippery. To clean stone, scrub with warm water and a non-caustic, sulphate-free detergent, then rinse with clean water.

Above, left: Slate's dark, moody tones give it a contemporary edge. Riven slate has a pleasingly rippled surface.

Above, centre: Slate tiles laid in a regular pattern look graphic.

Above, right: A bathroom floor of beachstones set in cement, which recalls Mediterranean courtyards.

Right: Japanese river-bed stones set in bleached concrete and ground smooth. The effect is terrazzo-like.

Hard tiles

Hand- or machine-made, glazed or unglazed, inlaid or relief textured, the diversity of hard tiles expresses a wealth of cultural traditions, from the Middle East to Mexico, from the Mediterranean to northern Europe. As practical as they are beautiful, there's a place for tiles in every home and tiles to complement every scheme.

The raw material of hard tiles is earth, a link preserved in the most evocative examples as warm colours and lively surface textures. Terracotta, that simple square of fired earth, has a history of use dating back thousands of years. Temporarily supplanted by the machine-made quarry tile in the mid-19th century, terracotta tiles have been rediscovered and valued for the way they age gracefully and display both their craft origin and the natural quality of their basic ingredient. Ceramic tiles come in a panoply of colours, finishes, shapes and sizes; encaustic tiles offer a wealth of decorative motifs to flooring schemes.

The practical appeal of the tile is that it offers the durability of a hard material in a format that is easy to work and handle. Like bricks, the human scale and domestic quality of hard tiles make them compatible with a wide range of applications. Unlike brick, however, the scope for decorative effects is truly vast. A tiled floor can be rustic or contemporary, provide a muted neutral background or take centre stage. From the soothing rhythmic

In most countries in the West, tiled floors are generally associated with hard-working practical areas of the home, principally kitchens, bathrooms and hallways. Aside from these obviously utilitarian applications, tiling can be very stylish in living rooms and dining rooms, particularly in warmer climates where their surface coolness is appreciated underfoot.

As with other flooring materials, keeping the scale sympathetic is important: larger tiles work better in generously proportioned rooms; small tiles generally suit more confined spaces. Floor tiling creates a sense of unity in rooms where the walls or work-surfaces are also tiled, although the same tiles may not be suitable for both applications. Floor tiles are thicker and heavier than wall tiles; smooth, high-glazed tiles with a glassy surface are not safe for use on the floor.

Like many other natural or near-natural types of flooring, the effect of tiling has been widely simulated in manmade materials such as vinyl and linoleum. In the past, some tile manufacturers responded to the competition by producing ranges that visually offered little more than their synthetic counterparts. Increasingly, however, there are signs of a renewed appreciation of the intrinsic appeal of tile, which has led to a revival of historic patterns and types, a greater demand for hand-made varieties and a growing interest in the diversity of local craft traditions.

quality of plain tiles to bold contrasts of colour and pattern, tiles have a liveliness that is frequently absent from wall-to-wall or sheet floor coverings.

Tiles share some of the attributes of other hard materials. Terracotta apart, most varieties are cold, but this can be resolved by under-floor heating. All are hard-wearing; ceramics especially. Maintenance is straightforward. On the down-side, hard tiles are noisy and can be slippery when wet. Their lack of resilience means they are tiring to stand on for long spells and anything dropped is likely to break.

The cost of floor tiles is as varied as the choice. Hand-made or reclaimed antique tiles command the top end of the range with machine-made tiles at the other end. A good-quality tiled floor costs about as much laid as the equivalent in hardwood. Laying, which often involves bedding in mortar, is professional work; cutting brittle tiles without damaging them to fit round obstacles is fiddly.

Far left: Patterned ceramic tiles are used to make a border within a border, adding decorative detail to a bathroom floor.
Left: Hard tiles are supremely practical in the kitchen.
Centre: Blue inset tiles sing out from a terracotta background.
Top: Mellow terracotta is homely and warm.
Above and right: Encaustic tiles are lively and decorative.

Terracotta, literally 'fired earth', are the oldest, and perhaps the most 'natural' of all hard tiles. They have an ancient history and were widely used in Roman times. The Moors introduced tile-making techniques to Spain between the 13th and 16th centuries, and the craft spread, to northern Europe and to Spanish territories in the New World. Handmade terracotta tiles are still produced on a small scale in many different areas of the world, using methods which have altered little over the centuries. In addition, there are antique reclaimed terracotta tiles on the market, some as old as two hundred years, as well as modern machine-made versions.

From one of the most basic of all ingredients and a relatively simple manufacturing process comes a surprisingly varied, characterful, product. Subtleties arise from the nature of clays and the variations in firing. Colours range from flinty grey to ochre, pink and traditional brick red, according to region. Provençal tiles are characterized by warm pink and yellow shades; ochre is typical of Tuscany. Several colours may be present in one tile as a result of blending different clays. The type of kiln used also contributes to the final effect — wood-fired kilns tend to produce livelier tiles than gas- or coal-fired ones — and even the way the kiln is stacked can affect the quality of the finish. Mexican terracotta, or Saltillo, for instance, has a rough and rustic look with flame marks enhancing the warm orange colour. The variety of colour and texture of handmade terracotta is matched by slight variations in thickness.

Left and above: Terracotta ages to a warm and mellow surface. The effect is so appealing there is a growing market in reclaimed or antique tiles.

Terracotta tiles are available in a variety of shapes, among them rectangles, hexagons and octagons, as well as the traditional 10-inch squares. Modern machine-made or extruded tiles are crisper and more contemporary looking than handmade varieties; they are sometimes supplied distressed to simulate the effect of a couple of centuries of wear. (A similar aged look can be achieved simply by laying tiles upside down.) However, the demand for authenticity is sufficient for a number of suppliers to specialize in sourcing antique terracotta 'pammets' from old farmhouse and manor floors. On occasion these pammets can be cleaned up to display a tilemaker's mark on the reverse.

Terracotta tiles are supremely at home in the kitchen, where their visual warmth is matched by an ability to retain more heat than other hard flooring materials, but they also provide a distinctive base for country-style living rooms, conservatories and any area connecting with the outdoors. These unglazed tiles are porous and require sealing which will deepen their colour. The characteristic glowing patina develops with time.

Quarry tiles are made from unrefined high-silica clay which is extruded into a mould, pressed and then burnt. 'Quarries' are the mass-production alternative to the handmade terracotta tile, and were first manufactured in Britain in the mid-19th century. Their popularity grew rapidly and large quantities were exported all over the world. Originally they were unglazed but vitrified versions are now available. For some, this eminently serviceable tile will always be Victorian, the flooring of kitchens and passageways of the English country rectory.

Quarry tiles come in the typically earthy shades of red, brown and buff, as well as in darker colours such as blue. Very dense and hard-wearing, they can nevertheless become abraded and pocked over time. Quarries are reasonably non-slip, and some varieties contain carborundum which provides a fully non-slip surface. However, all quarries are much colder (although cheaper) than terracotta. Fully vitrified quarries can be used outdoors on terraces and porches. Quarries are available in a range of sizes and thicknesses; square tiles are standard.

Above left: Quarry tiles are quintessentially Victorian, the perfect complement to a country kitchen or farmhouse hallway.

Above centre: Large handmade modern terracotta brings a natural quality to a contemporary floor. The patina develops with time.

Above right: Old terracotta tiling is a floor to treasure. Colour, texture and shape of tiles vary from region to region.

Right: Cream and red terracotta set in a chequerboard pattern makes a sympathetic hearth in a country farmhouse.

Ceramic tiles are regularly dimensioned and coloured, which gives them a crisp, contemporary appearance that works well with modern interiors. Different sizes of tile create different textural effects.

Above and above centre: Very small tiles make a busy grid which is soft-looking, like mosaic. Borders or contrasting bands of colour lend definition and interest.

Left: Larger tiles are more expansive and spacious in effect.

Ceramic tiles are made from refined clay, which is ground, then pressed into moulds under great pressure before being fired at very high temperatures. The result is an exceptionally durable tile which is very regular in dimension and colouring. This precision gives the tiles a crisp, contemporary look, accentuated by the fact that the tiling can be more closely spaced since sizes and shapes are so accurate. Understandably, ceramic tiles are most at home in clean-lined modern interiors or wherever such regularity is an asset.

Cold, hard and wear-resistant, ceramic tiles make an ungiving and somewhat tiring floor but one that is impervious to water and most stains. Fully vitrified versions are also frost-resistant. In general, ceramic tiles are slightly slippery, but non-slip versions containing silicon carbide are available, and ones with ribbed, ridged or studded textures give a better grip underfoot. Glazed tiles are less good at resisting wear and matt tiles can wear down to a smoother surface. Ceramic tiles can be expensive and are generally heavy.

The beauty of ceramics is the astonishing choice of colours, shapes, patterns and textures. The familiar rustic palette of earthy shades is complemented by a whole range of more vivid colours in solid, shaded or variegated forms, and tones of white, black and grey. Patterns are equally varied, from the verve and individuality of hand-painted decoration to machine-made glazed, embossed or relief designs. Unlike terracotta, ceramic tiles do not acquire any discernible patina with use, but the fact that they are relatively unchanging can be welcome where a fresh, clean look is desired.

Encaustic tiles have a longer history than their common link with the Victorian era suggests. A technique of inlaying tiles with designs was developed by monks in medieval Britain and the results, featuring Christian, heraldic and geometric motifs, can be seen in many old churches and religious buildings. The Victorian passion for the Gothic is evident in the 19th-century version of encaustic tiles. Their mass-production was developed by Herbert Minton in mid-century and by the latter decades encaustic tiles had become the standard flooring for hallways and porches in terraced houses.

The term 'encaustic' refers to the process of firing, but it is the decoration that is so distinctive. Encaustic tiles are patterned through and through. The basic material (clay or stoneware), still in a semi-liquid state, is inlaid with a pattern in a different coloured clay, not superficially decorated or glazed. Some versions are a mixture of stone and powdered marble coloured with oxides, and are not fired.

Encaustic floors went out of fashion in the early 20th century, but many were simply covered over. Reclaiming the floor restores an element of architectural character to a period house. (Replacement tiles, new or secondhand, in original Victorian designs can be found.) Encaustic tiles are uniformly smooth, matt and cool. They naturally discolour slightly, with clear whites dulling to off-white. They stain easily before they are sealed and should be handled with care.

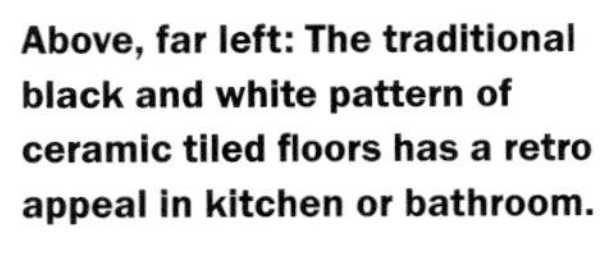

Above, far left: The traditional black and white pattern of ceramic tiled floors has a retro appeal in kitchen or bathroom.

Above, left: A classic chequerboard design in a homely colourway makes a practical kitchen floor.

Top: New cement tiles have been given an aged finish.

Above: A tumbling block pattern has movement and vitality.

Far left: Encaustic tiles are largely associated with the Victorian and Edwardian periods, although their history dates back to medieval times. Gothic designs are typical, but new versions display more organic or pictorial themes.

Centre left: In Victorian houses, encaustic tiling is a typical flooring for entrances and halls.

Left: Classic white squares with diamond insets, in ceramic tile.

Patterning

Tiling offers unlimited scope for creating patterns, often very simply. All tiled floors have an inherent pattern, arising from the grid in which they are laid. The dynamic of this composition depends on the size and shape of the tiles, and whether they are laid straight, with staggered joints, or on the diagonal.

Classic tile patterns range from octagonals or squares inset with keystones to basketweave, herringbone and diamond. The effect can be enhanced by varying the tile type, colour or pattern: small pictorial tiles can make a charming contrast to a plain background. Some patterns, especially those using lozenge-shape tiles, have a Hispanic flavour; others recall medievalist designs. Complex patterns such as 'tumbling block', a three-dimensional optical effect seen in quilting designs, can be created with tiles of different colours or tones. In general, the busier the pattern, the smaller the floor area should be. Complex designs are also effective as centrepieces within a larger expanse of plainer tiling. Many suppliers have a custom design service to help plan and execute special patterns.

Borders give a finished look. If the room is irregularly shaped, it may be better for the border to be set in from the wall to make an even outline rather than follow the contours of the room. Joints between border tiles should be staggered rather than aligned with the main tiling joints. Special coving or skirting tiles are available which make a neat seam with the base of the wall; skirting tiles in contrasting colours or designs give the whole floor a lift.

Laying

All types of hard tile require a rigid, level base or the tiles will crack. Heavier ones, such as ceramic, may place too great a load on a wooden subfloor: check with an expert if you are at all in doubt. A dampproof membrane may be needed. Tiles are either bedded in mortar or stuck with commercial adhesive, which should be waterproof if you are tiling a kitchen, bathroom, conservatory or utility area. Mix a commercial additive into the adhesive if tiles are to be laid over suspended wood floors to promote flexibility. Terracotta tiles should be stuck with adhesive, not bedded in mortar, to reduce 'efflorescence' – the formation of salt deposits on the surface.

Lay the tiles dry first, to minimize the number that have to be cut to fit, and so that grout widths can be adjusted if the tiles vary slightly. Dry laying is also vital if you are incorporating a border or a pattern. Glazed or decorative inset tiles within an area of plain tiling must be countersunk or they will abrade. Laying plastic, felt, or building paper over the subfloor, using a semi-dry mortar mix, and bedding in latex mortar allow a little movement, so tiles do not arch up. Movement joints, filled with elastic sealant, should also be incorporated, usually around the perimeter of the floor, over structural elements such as beams and at recommended intervals.

Terrazzo

Terrazzo is an aggregate of marble or granite chippings mixed with concrete or cement to form a beautiful yet exceptionally hard-wearing floor. Excellent performance in areas of heavy traffic combined with classic good looks has won terrazzo widespread use in commercial interiors, from cafes and hotel lobbies to retail outlets. Familiarity with such applications may give terrazzo a slightly impersonal quality in a domestic setting, but it can make an elegant, even sumptuous floor for hallways and other hard-working areas of the home – and in warm climates it is often favoured throughout the home as a delightfully cool treatment for floors.

Terrazzo tiles and slabs are laid in much the same way as other hard tiles, with cement slurry applied to the backs before laying. Tiles are normally laid in groups separated by metal dividing strips, and they may be ground after laying.

Terrazzo is more expensive than most types of hard tile and only slightly cheaper than the best quality natural stone. In practical terms terrazzo is cold, noisy, hard, waterproof and virtually indestructible. Hydraulically pressed tiles are extremely durable. The smooth surface is reasonably non-slip, unless it is wet, washed with soap, or polished.

Terrazzo has a much longer history than you might think. Hand-made terrazzo tiles have been produced in Mediterranean countries for over a century, where they remain a popular flooring material indoors and out. They are available in a huge range of colours, and vary in appearance from the standard mottled mosaic of chippings to crisp geometric designs which lend themselves particularly well to modern settings, a pleasing complement to chrome fittings, pale wood and glass.

There are several formats. Terrazzo can be mixed and laid in situ, or laid as tiles or slabs. In all cases, laying is professional work. In situ terrazzo is trowelled onto a solid concrete base or onto screed bonded to a concrete base to form panels within brass or zinc dividing strips. It is then ground to produce a smooth finish, washed, filled with cement paste, cured and polished.

Maintenance

General maintenance is straightforward. Terrazzo is best washed with warm water and a little scouring powder; soap makes the surface too slippery. Water-based seals will protect against staining. Avoid polish containing wax.

Above, left: Terrazzo is sleek and contemporary. The link with public spaces has given its crossover into the home an exciting edge.

Above, centre: Terrazzo owes its luxurious appearance to ingredients such as marble chippings, coloured glass and stones, ground smooth.

Above, right: Patterns and motifs can be achieved using contrasting colours. Laid in situ, terrazzo can be expensive.

Right: The smaller scale of these terrazzo tiles suits the domestic setting. The material is simple to maintain and very robust, making it ideal for hardworking areas.

Mosaic

Mosaic is true floor-level art. Small cubes bedded in mortar in decorative or geometric designs give mosaic an irresistible intricacy and delicacy. The scale of individual pieces and the variations of light catching on their surfaces create a gentle, almost blurred effect. Mosaic floors are hard, but their appearance is soft.

Any size floor can be covered with mosaic, although in practice it tends to be used on a smaller scale, in bathrooms, in hallways or as decorative panels inset within a larger tiled or flagged floor. The technique is exceptionally labour-intensive, which also accounts for the expense of a commissioned mosaic floor. Simple geometric mosaics are not impossible for the amateur to tackle, but for more complex patterns, you need a mosaic artist to design and carry out the work. A good artist will be able to show you examples of other projects and produce an original design to suit a particular location.

Mosaic flooring is generally composed of cubes or *tesserae* of marble, stone, terracotta or unglazed through-coloured ceramic. Different materials can be combined within the same mosaic. Vitreous glass mosaic is not suitable for flooring because it fractures easily under point loads and can be too slippery. Honed marble is generally used in preference to highly polished varieties. Marble and other stone *tesserae* tend to vary in thickness, making them harder to install than ceramic mosaic, which is evenly sized.

Mosaic shares many of the characteristics of other hard floors: durable, noisy and fairly cold. However, the natural key supplied by the myriad grouting joints means that mosaic is far less slippery than floors made of larger tiles or slabs of the same material.

Below: Mosaic pavements featured in buildings throughout the Roman Empire: current designs are often inspired by antique examples.

Centre: A Greek key pattern adds a classical flavour to this mosaic.

Centre, right: A mosaic surface shimmers as light catches on the tiny cubes of stone and tile.

Far right: Geometric patterning is easier for amateurs to create than full-scale pictorial designs.

The grouting also contributes far more positively to the overall effect. Intensely coloured mosaic is enhanced by dark grouting; light grouting tends to dull strong colours and to make designs look more fragmented. Light grouting makes pale mosaic look more seamless; dark grouting gives a more drawn or graphic effect.

Laying

Mosaic can be heavy. It is best laid on a concrete sub-floor, but it can be laid on a suspended timber floor covered with a layer of marine plywood to provide a stable, inflexible base. A deep bedding of mortar will take up any discrepancies of thickness. Ceramic mosaic can be stuck in thin bed adhesive over a flat screed. Check with a surveyor if there is any doubt about load-bearing capacity.

There are two main ways to create a mosaic floor. One is the direct method, which involves sticking in the individual pieces one by one. It is best used only for small areas, as it is time-consuming, and restricted to simple designs, so that there is less room for error. The most practical is the reverse method. The mosaic is first stuck to paper with water-soluble glue, with the finished surface face down. The back of the mosaic is then grouted or covered with a sand and cement slurry to fill in the gaps and the mosaic pressed, right side up, on the floor. The paper is wetted to dissolve the glue and peeled away, and the mosaic is washed to remove all traces of glue and left to 'go off', or cure, before final grouting.

The great advantage of the reverse method is that it minimizes installation time, since the lengthy process of cutting and piecing the design can take place away from site before the mosaic is slotted together in finished sections. As the complete floor is built up from sections, care must also be taken to fit these together properly so that joins are not visible.

Maintenance

Mosaic, particularly marble mosaic, must not be highly polished after it is laid or the natural key provided by the difference in texture between the mosaic and grouting will be lost and the surface will become dangerously slippery. Wash with a neutral detergent.

Above, left: A concrete floor at its most basic: defiantly brutal.

Above, right and right: Concrete can be treated to alter its colour and texture. Otherwise, as these examples show, concrete can be left in its 'natural' state. Concrete can be visually soft, with some textured finishes resembling suede. Raw concrete surfaces should be sealed to prevent the surface from dusting.

Left: Pebbles set in a concrete floor give the impression of 'tiles'.

Far right: Concrete's appeal as a modern floor is its uncompromising and strong, sculptural quality.

Concrete

Concrete is the ultimate workhorse floor, utilitarian, basic, even a little brutal. Its popular image as a soulless, harsh material is not undeserved in its raw state, but screeded and painted, waxed, textured or coated it can be transformed into a sleek, sophisticated finish of considerable stylistic merit.

Concrete is a blend of cement, aggregate and sand. It is available in slabs or tiles, or mixed and laid directly on site. Its main domestic use is as a sub-floor for other, generally heavy coverings, or as practical but unlovely flooring in utility areas, cellars and garages. There is, however, a variety of ways in which concrete can be lifted out of the ordinary. Some require the rough concrete base to be covered with a smoother sand and cement screed. For a finish not unlike terrazzo, special aggregate can be incorporated in the mix and the laid floor then ground and polished after it has set. A quirky effect can be achieved by scoring patterns or embedding pebbles while the concrete is still wet, then sealing and waxing.

Concrete must be sealed to prevent the surface from dusting. Special concrete floor paints give an attractive and durable finish available in a range of colours. Concrete can also be coloured with the same pigments used for plaster for a soft, matt look. Several coats of acrylic or epoxy resin result in a hard reflective finish which is very tough and resistant to chemicals. Other types of resin or aggregate toppings are textured and non-slip. Most of these finishes are designed principally for industrial or commercial applications, and, like the floor itself, should be laid by a professional for the best results. Once laid, though, concrete needs little more maintenance than scrubbing with hot water and detergent.

Painted a glossy neutral tone, concrete achieves a high level of sophistication. The reflective sheen adds spaciousness and light.

Concrete's tough aesthetic works well with the clean lines and metal finishes of an industrial-style kitchen.

Metal

Another migrant from commercial or industrial settings, metal flooring brings a sleek contemporary edge to the interior. The hard shiny surface, with its aeronautical associations, makes a bold statement in a modern space. It is available in sheet and tile form, and is normally textured with raised patterns, such as 'treadplate', to provide a non-slip surface. Aluminium and galvanized steel are the most commonly used metals; zinc is too soft. Aluminium is up to three times lighter than galvanized steel and does not rust.

Metal sheet can be laid over a level timber or concrete sub-floor and may be either bonded with adhesive or screwed and plugged in place. If the floor is to be mechanically secured, a little adhesive should also be applied to take the flex out of the floor and prevent

it from rattling when you walk across it. Even with this precaution, the surface will be noisy and cold underfoot. Metal flooring can be sealed and is best maintained by washing with water and detergent.

Glass

Glass flooring is the ultimate in drama. At upper levels, as a mezzanine walkway, glass has a 'look-no-hands' quality which adds an exciting dimension to the interior and allows clear views through a normally vertically divided space. Its application in domestic contexts is not widespread, though – and it is expensive.

The type of glass suitable for flooring is thick annealed float glass, rather than toughened glass. Every glass floor should be individually specified to meet loading requirements, but in most domestic situations adequate strength is provided by a top layer 19mm thick laminated to a 10mm base layer. To keep weight and

Above: Metal and glass heighten the drama of stairways.

Centre: Metal laid in tile form makes an unusual floor for a dining area.

Top, right: The ultimate glass flooring: display cases topped with glass laminate.

Bottom, right: Glass needs friction bars to be a safe walkway.

manoeuvrability within reasonable limits, it is usual to work in metre-square panels. Glass walkways must be supported on all four edges by a timber or metal frame that entirely encloses the thickness of the glass. As glass cannot be butted up against another hard material, edge cushioning in the form of neoprene rubber must be included. The rubber must be hard enough to withstand compression. To counter the extreme slipperiness of the floor, it is usually necessary to have friction bars sandblasted at intervals across the surface.

Resin

Not strictly speaking a hard flooring material, rather a treatment or topping applied to a solid base, resin flooring is increasingly popular in contemporary homes. Epoxy resin flooring was originally developed for commercial and industrial applications. Smooth, sleek and available in a wide range of colours and textures, it is often preferred by home-owners over similar materials, such as polished concrete or poured rubber, because it is much easier to get right first time. Poured rubber floors are notorious for failing and it can be difficult to get a perfectly smooth and level concrete base.

Resin offers the same degree of seamlessness with a similar contemporary edge. Like many industrial products, it is designed to take a great deal of punishment and to be easy to maintain. Resin resists chemical attack and heat and is antistatic and very durable. There are different specifications, including self-smoothing resins, water-based resins and those that are solvent-free. Resin requires specialist application, but is simply mixed in situ, rollered on and left to cure. At least two coats are applied, sometimes more, depending on the required thickness. The usual base is concrete, although it can also be applied over other substrates such as wood.

Far left: Resin floors come in a range of colours, which makes it easy to match existing decor.

Left: A blue resin floor with glass brick insets.

Below: Resin flooring is particularly appropriate in kitchens, as it is very hygienic.

Right: Glossy sky-blue resin extends throughout this contemporary apartment.

Wooden floors

Waxed oak parquet, dark stained boards, pale ash strip or honey-coloured pine, wood is one of the most beautiful and accommodating of all flooring materials. It is also a material with which we have an instinctive rapport, born out of centuries of familiarity. Wood has found a role in almost every aspect of the construction, detailing, decoration and furnishing of houses, from doors to window frames, from skirting boards to balustrades, from chests of drawers and kitchen tables to – floors.

Very few other materials are so versatile or so commonplace. But this familiarity has not bred contempt or even indifference; wood has proved remarkably immune to the swings of interior design or architectural fashion, and is equally at home in the penthouse suite and the rural farmhouse.

The eternal appeal of wood owes much to the fact that it derives from a natural, living source. The pattern of growth evident in knots and grain give wood an essential vitality and variety which makes it easy to live with in large doses. No two boards are alike, no two species are alike. Wood, in fact, is more of a family of materials with related but different characteristics. Like all natural materials, it ages well and, with sympathetic care, lasts well, acquiring a depth of character that money cannot buy or manufacturers simulate.

The variety of tone or colour, texture and density of bare wood is complemented by a range of formats, from sheets, strips and boards to mosaic tiles, blocks or parquet. The rhythm and scale of each format have a large impact on the overall appearance of the

Above: Plain, 'unfinished' floorboards provide a counterpoint to furnishings with a period flavour. The pale tone of old oak also increases the sense of light and spaciousness.

Above, left: This parquet floor, laid in a herringbone pattern, has been stained a dark colour and highly polished for a chic, elegant effect.

Left: Hardwood flooring is a classic contemporary surface. Beech is currently one of the more popular woods.

floor. Small-scale tiles and blocks have a busier, more enclosing look. Wide boards look rugged and rustic, narrow strip flooring has a seamless quality. Then there is the decorative dimension. Wood can be simply sealed and left to speak for itself or treated as a blank canvas and bleached, stained, painted and stencilled. In addition, it provides an attractive setting for rugs and matting.

Neither hard nor soft but somewhere in between, a wood floor offers many practical advantages, the best, in some ways, of both worlds. It is cooler and 'breezier' than carpet but warmer than stone and hard tiles. Most wooden floors, unless they are laid over concrete, have plenty of resilience or 'give', which makes them comfortable to walk and stand on and less liable to cause breakages when something is dropped.

Above, right: A broad herringbone parquet creates a sense of expansiveness in a narrow room. The matt waxed finish is subtle and practical.

Far right: The rich hues of an elm floor can look just as well in an uncompromisingly contemporary context as in a more traditional setting.

Left: New French oak floorboards add just enough warmth to what could otherwise be an austere colour scheme.

Wood is not brutally noisy but it is no guarantee of a quiet life. A suspended timber floor in a room with few soft furnishings can add to the echoing quality, while expanses of bare boards at upper levels can be quite rackety, especially for those in rooms below. If you live in a flat, consider your downstairs neighbours before opting for a bare wood floor, otherwise good relations could become strained. It is, however, easy to muffle the sound with a layer of rugs or matting on those areas of the floor which take most of the daily traffic and still enjoy the beauty of exposed wood around the perimeter; in extreme circumstances, underfloor soundproofing can radically decrease noise levels. But for many people, the odd creaking board or squeak of parquet is all part of the evocative appeal of wood.

Wood comes in most price ranges. New timber, especially new hardwood, is expensive, but it will easily outlive, say, a top of the range carpet. A solid oak floor should last for generations. In the mid-range, wood strip and block flooring are available from major outlets, while cheapest of all are manufactured sheets such as plywood, which can make a surprisingly stylish floor. A veneered flooring is cheaper than the equivalent solid wood version, but correspondingly it has a much more limited life, since the veneer is usually too thin to be stripped and refinished once it has worn. Renovated boards generally demand more hard labour than hard cash.

Above far left: Inlay of cherry fruitwood makes a sumptuous finishing touch.

Above left: Edging pine with a darker inset adds a neat detail.

Above: Herringbone parquet makes a lively contrast to the gridded pattern of the panelled walls.

Left: Gleaming boards match the modern feel of metal fittings.

Wood is more demanding than a hard floor in terms of care and maintenance. It is by no means proof against all damage, even when properly sealed. Grit tracked in on the soles of shoes wears down seals and lets in moisture and dirt; stiletto heels and narrow furniture legs are incredibly destructive. Even under normal circumstances, seals and finishes need to be renewed from time to time. Really bad patches of wear may mean that the entire floor will need to be refinished. But the result is a long life of good looks.

There are few areas of the home where a wood floor would not be welcome, either stylistically or practically. Functional as well as beautiful, wood works equally well in kitchens as in living rooms, dining rooms, bedrooms and playrooms. Its susceptibility to changes in humidity, however, means its use is not recommended where exposure to water and steam — in poorly ventilated bathrooms, for example, or conservatories — would inevitably compromise its durability. All-weather decking can extend wooden flooring right out onto the veranda, terrace and beyond.

Right: Wood can create a wide range of effects very economically. These plywood tiles inset with dark squares provide a variation on a classic flooring pattern. Well sealed, plywood makes a surprisingly durable floor for a fraction of the cost of hardwood.

Below: Parquet laid in a basketweave pattern complements contemporary decoration.

Below right: Boards in different woods create a lively surface.

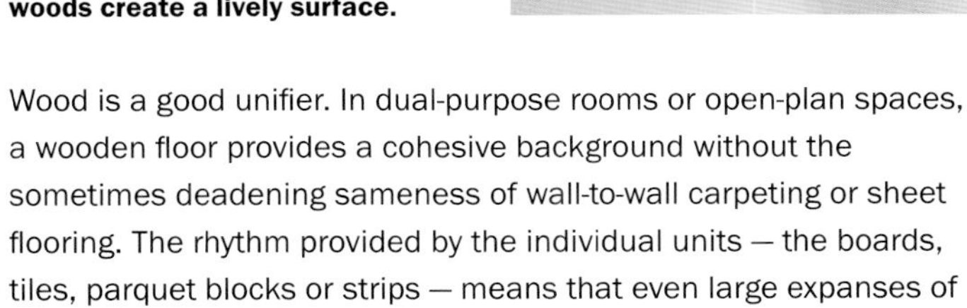

Wood is a good unifier. In dual-purpose rooms or open-plan spaces, a wooden floor provides a cohesive background without the sometimes deadening sameness of wall-to-wall carpeting or sheet flooring. The rhythm provided by the individual units – the boards, tiles, parquet blocks or strips – means that even large expanses of wood retain a homely quality.

After centuries of use we show no signs of tiring of this classic material. In fact, as interest grows in natural sources and products, wood has acquired a heightened level of appreciation and has become a more popular choice of flooring than ever. With this shift in emphasis has come an equivalent concern to ensure that timber used in the home should be from sustainably managed estates and other renewable sources that do not lay waste to irreplaceable rainforests or indigenous woodlands.

New wood

Classic and contemporary, new wood is one of the most elegant of flooring materials. Physically warm, aesthetically cool – it provides sheer quality underfoot. But stylishness does not rule out more homely virtues. The best wood floors mellow to a comfortable patina which is as pleasing as the finish on a cherished piece of furniture.

New wood encompasses a wide range of species and formats, with equivalent variations in price, durability, suitability and appearance. Solid hardwood floors are expensive, require professional laying and informed selection as to species – but they are extravagantly beautiful and exceptionally durable. At the other end of the spectrum are ready-made wood floors which come with a mass-market price tag and can be laid by the skilled amateur. But, as ever, you get what you pay for and the downside to these more affordable alternatives is a relatively short life which often cannot be prolonged by refinishing. Nevertheless, with proper care, even the cheapest wood offers the pleasure of contact with a natural material, and the appealing blend of associations that wood inspires.

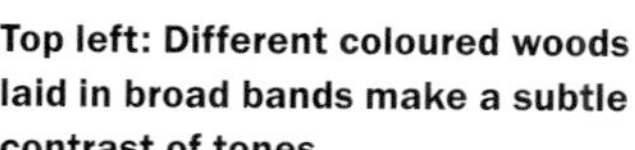

Top left: Different coloured woods laid in broad bands make a subtle contrast of tones.

Top right: Hardwood laid on a solid stairway adds texture and warmth.

Above and above centre: Strips and blocks provide rhythm and interest.

Right: Hardwood laid herringbone fashion accentuates the sweeping transition from area to area.

The densest woods are the most resistant and hard-wearing. Some timber is strong and hard enough to function as industrial flooring, able to withstand the heavy loads of factory machinery or the incessant traffic of a school corridor. Needless to say, such performance levels are never required in the home, but different types of wood do vary in their practical characteristics and some are better in certain contexts than others.

Durability is also a function of how the timber is worked. The toughest is end-grain, followed by quarter-sawn, where the timber is cut radially for more even grain pattern and greater stability. Plain sawn timber wears less well and may be less attractive (plain sawn softwood, for example, is very knotty), but it is cheaper and more readily available. Veneered boards or tiles are least durable of all.

Left: Wood flooring is a good unifier in open-plan spaces, softening clean modern lines with its warm, natural tones.

Right: The pale tones of oak make a perfect surface in a modern house by the sea, emphasizing its dramatic light-filled vistas.

Below: New hardwood flooring, such as this oak strip, has the potential to improve with age, amply repaying the initial high investment.

Left: The even tones of beech make it a popular, easy-to-live with, choice for modern floors. It is strong and very hard-wearing. Press-dried beech can be harder than oak, and it is less expensive.

Types and characteristics

Timber species broadly fall into two categories: softwoods and hardwoods. Softwoods, as their name implies, are generally less durable than hardwoods. They are also widely available and cheap. The two main softwoods are pine and deal. Their principal uses are as replacement timber for original softwood floors which have become damaged beyond salvation and in new construction. Hardwoods are more expensive than softwoods and display an incredible variety of colour, pattern and texture. New wood floors are generally manufactured from hardwood, which may be either solid or in veneer form.

In recent decades many hardwood species, particularly from tropical regions, have become endangered. Uncontrolled deforestation has destroyed vast areas of rainforest around the world, with disastrous ecological consequences. Some of the species most at risk include afromosia, iroko, keruing, mahogany, sapele, utile and teak, all of which have flooring applications.

It is up to you, the consumer, to be responsible when choosing wood. In general, it is advisable to avoid buying any imported tropical timber unless it has been produced on a managed plantation and is certified as such by the Forest Stewardship Council. This international body keeps track of properly managed forestry projects around the world which provide sustainable sources of times as well as livelihoods for local communities.

New woods suitable for flooring from temperate forests in Europe and North America include:

Ash Pale, tough with a coarse texture and straight grain. Good for general use.
Beech Attractive, light toned and very strong and durable. Widely used in block flooring. If press-dried can be harder than oak.
Birch Pale, finely textured. Not strong. Commonly used as plywood.
Cherry Fine grained, rich warm colour with slight pinkish tinge.
Chestnut Exceptionally strong and durable.
Elm High water-resistance and dark rich colour. Very strong.
Lime Pale, straight grained and fine textured. Good for general use.
Maple Excellent flooring timber. Will withstand heavy traffic and often used as flooring in museums, schools and ballrooms. Reddish tinge.
Oak (American, English and French varieties). The classic flooring timber. Coarse grained. Strong and durable. Resists water, rot and pests.
Pine Light softwood, available in various grades. Honey-toned when sealed. Less durable than hardwood, but very economical. Must be treated against rot and woodworm before use.
Sycamore Light toned, but takes staining well. General use.
Walnut Matures to rich colour. Wavy grain.

Of these, ash, beech, maple and oak are the most common hardwoods used in flooring production. Many flooring manufacturers produce boards in a variety of finishes and colours, so the same species may be offered in its natural tone, bleached, oiled, lacquered, coloured or stained to resemble a tropical wood. Some strip flooring is treated with acrylic hardener to increase its durability. In addition, boards may be supplied untreated so that you can apply your own environment-friendly finishes.

All wood must be correctly 'seasoned' before use. Seasoning is the process by which newly cut wood gradually loses moisture until it reaches an equilibrium with the atmosphere, shrinking as it does so. If you laid green timber boards in a centrally heated room, before long the wood would have shrunk considerably, possibly warping and splitting in the process. Similarly, wood which has dried out too much will take up damp in a humid location and swell. Flooring timber is generally kiln-dried to a moisture content of 10% or less. Major flooring manufacturers supply wood that is ready to use; in other cases, you may need to keep the wood in the room in which it will be laid for a period of between 10 days and two months in order for it to acclimatize. The amount and type of heating you have in your home can be critical. If you like the thermostat turned up or have underfloor heating, the wood will need to have a lower moisture content.

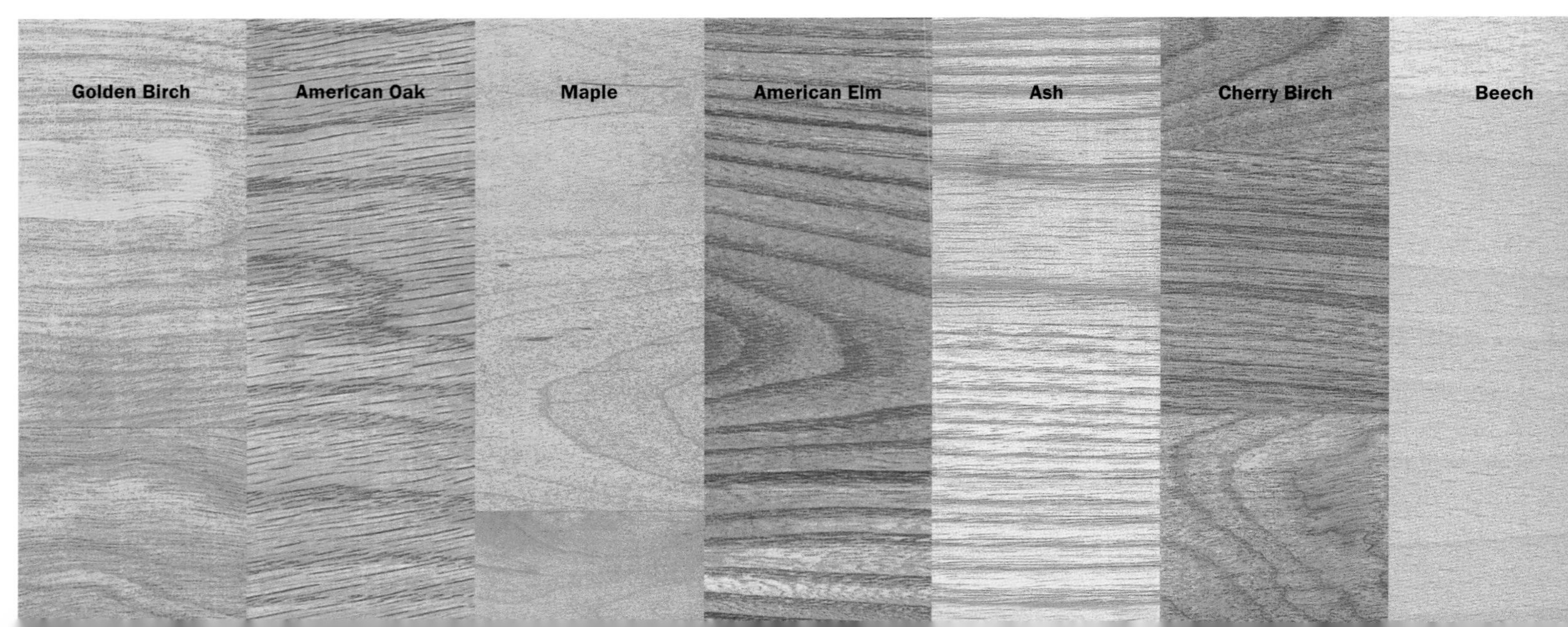

Below left: Pirainah pine laid herringbone fashion and edged with a darker strip inlay makes a classic entrance.

Centre: New wood softens the otherwise stark features of this small room. Narrow strips work best in a restricted space.

Below right: Humble pine gains distinction laid in wide boards, sanded and sealed for a hard finish.

Bottom: Parquet flooring makes a sleek contrast to a rugged exposed concrete ceiling and beams in a converted factory.

The terminology for wood flooring can be confusing. 'Parquet' is sometimes used synonymously with strip flooring; 'block' and 'parquet' are also used interchangeably. 'Boards' generally refer to softwood floorboards; 'woodstrip' can mean anything from a length of hardwood of similar dimensions and thickness to thin overlay.

Parquet or block flooring has a long history of use, reaching the height of artistry in 17th- and 18th-century French châteaux. Many châteaux boasted their own individual designs, created in situ and resembling the finest inlaid furniture. Composed of small strips or blocks of hardwood, most typically oak, parquet is laid in patterns which range from the familiar herringbone to more intricate designs featuring interlocking lozenge shapes or woven basketweave.

Modern parquet is supplied tongued and grooved or dowelled and is made from a variety of different hardwoods. Solid wood parquet can be refinished, providing the tongues or dowels are not too near the surface.

Other forms of parquet or block consist of solid elements glued together, or a hardwood veneer over a softwood base. Some of these products can be loose-laid; others laid on a dry, existing floor, provided it is perfectly flat. Solid parquet can be refinished. For an *ancien régime* look, there are also suppliers specializing in reclaimed parquet.

Wood mosaic Essentially these are wooden floor tiles, consisting of thin strips or 'fingers' of wood glued together to form squares and bonded to a base of scrim, felt or adhesive backing. Some are faced with paper or a membrane which is removed after laying. A tile or mosaic panel usually consists of four squares, set at right angles to each other to form a basketweave pattern. The size, thickness and type of wood vary.

Panels are usually tongued and grooved and may be adhesive-backed or stuck in place. Mosaic has good resistance and is less susceptible to movement due to changes in moisture levels. It does have a rather dated appearance, however, and lacks the grandeur of solid wood flooring or the elegance of parquet.

Woodstrip Hardwood strip varies in thickness, length and width. Generally the larger the dimensions the more expensive the wood. Narrow woodstrip flooring can look a little institutional, uncomfortably reminiscent of the school gymnasium; wide long boards are more luxurious. Many manufactured floors consist of strips glued together to resemble boards.

Thick planks or strip floors can be laid over joists like floorboards; thinner strip flooring must be laid over a solid sub-floor or a dry, flat existing floor. Both types are usually supplied tongued and grooved

Opposite page: Woodblock flooring made of end-grain, the hardest and most durable, makes a practical surface in a dining area.

Top: Pale wood has an appealing timeless quality.

Above: Parquet is made from a variety of hardwoods, often oak.

Left: The rhythm of wood mosaic is complemented by natural fibre rugs, which also lessen the noise of footsteps on bare wood.

so they fit together without leaving any gaps. One international supplier offers thick strip flooring which can be laid like ship's decking; the boards are tapped together, with coloured neoprene in the joins compensating for any movement, and reducing slipperiness.

Thickness is a critical factor, and ranges from about 8mm to 25mm. Woodstrip of medium thickness can be refinished if it is solid wood and the tongue is well below the surface, but woodstrip which consists of a layer of thin hardwood veneer over a softwood or composition wood base cannot be sanded.

Manufactured strip flooring or overlay is designed to be easy to install and maintain, as well as economical: at the cost, however, of a certain degree of authenticity. Some laminated strip flooring has only the merest skin of wood facing; the effect, inevitably, is rather synthetic. High-pressure laminates can withstand heavy-duty traffic, spike heels and cigarette burns.

Veneered wood floors backed with cork for extra resilience and comfort are also available. Strips as thin as 4mm can be glued to sub-floors or, twice as thick, loose-laid as a floating floor over an existing one. Like laminates, they are very resistant to damage by heavy furniture or high heels and require no subsequent varnishing. The cork cushioning makes them ideal for laying over concrete, and they are also quiet, non-slip, anti-static and hypo-allergenic. Purists might quibble that such floors retain a manufactured quality which lacks the depth of character of solid wood.

Left: Woodchip is not generally recommended for flooring purposes, but it makes a lively temporary covering.

Far right: Types of plywood suitable for flooring include birch- or maple-faced, and those that are birch throughout.

Laying

Laying a new wood floor is usually professional work. Good hardwood represents a significant investment which could all too easily be spoiled by a botched amateur job; intricate parquet patterns demand skill. The exception is manufactured strip flooring, which is fairly straightforward to install (see pages 174–75).

Sub-floor requirements vary. Boards and thick woodstrip can be laid on joists or battens, which means that piping and cabling can be run under the floor. Maximum spans are dictated by the thickness of the wood. Boards and thick woodstrip can also be used over under-floor heating. Thinner woodstrip, mosaic and some types of parquet can be laid over any dry, flat floor. Existing floorboards should be covered with hardboard or cardboard. Concrete may need to be covered with a damp-proof membrane or resilient underlay. Solid parquet is usually laid on screeded concrete.

Left: Plywood may not be as long-lived as hardwood, but it is so economical that reflooring is perfectly affordable. It should always be well sealed and is best not laid in areas of heavy traffic.

Plywood

Plywood has long been a by-word for the flimsy and makeshift, a manufactured wood fit only for temporary or utilitarian uses. But in recent years the material has been rediscovered: cheap and nasty has undergone a transformation as cheap chic. What was once considered appropriate only as a base for sub-floors has emerged as a stylish covering in its own right.

Early modernists, such as Le Corbusier and Mies van der Rohe, were among the first to appreciate the suave potential of this machine-made wood; pre-war, plywood was used for flooring, among many other applications. During the 1950s and 1960s, the material was more or less eclipsed by other manufactured wooden flooring products, as well as wood-effect simulations in vinyl.

The latest reincarnation of plywood has come about as the result of a renewed appreciation of its practical and aesthetic qualities. Plywood is modern, natural and agreeably utilitarian, as opposed to strip flooring, which can sometimes take on more glitzy commercial overtones.

Although plywood is manmade, it is wholly wood — not synthetic in any sense. It is composed of thin sheets of wood glued together under pressure and its strength derives from the fact that sheets are layered at right angles to each other. Composition varies, with birch being the most common ingredient. Birch, which is not a particularly strong wood, gains strength in this format. Types of plywood which are suitable for flooring include those that are birch-faced or maple-faced on a softwood base and those that are solid birch throughout. No plywood is long-lasting. The cheapest, veneered or faced, types will wear quickly; the solid birch ones have a slightly longer life and can be gently sanded.

Plywood panels are widely available; the best sources are timber merchants. Thicknesses of up to 20mm can be supplied cut into squares and tongued and grooved. Provided the size of panels is not too unwieldy, laying a plywood floor can be successfully tackled by a competent amateur. Plywood is sufficiently flexible to accommodate slight unevenness in existing sub-floors. It can be laid on boards or joists if it is thick enough or bedded in bitumen on screened concrete. Make sure the grain faces in the same direction. Sand lightly and seal thoroughly with a good lacquer to bring out the warmth and lustre. Plywood can also be stained or painted before sealing.

Above and above left: Original wood floors have immense character and are the perfect complement to period or contemporary decor. This floor is composed of unpolished wood blocks laid in an intricate pattern.

Left: The textural character of these walnut boards arises from the fact that they were hewn with an adze, rather than machine sawn.

Below left: Maple is a beautiful flooring material, often found in schools and dancehalls.

Right: To create this 'period' floor, old oak boards were laid at varying heights and spacing, then scratched and bleached for a distressed surface, rather than sanded smooth.

Old wood

In the same way as wall-to-wall carpeting defined the new comfort and informality of the Western lifestyle in the 1950s and 1960s – along with the three-piece suite, fitted kitchen and open-plan arrangement – stripped floorboards have become a contemporary decorating standard, and for very good reasons. Renovated floorboards are cheap and characterful, versatile and practical and make an easy match with any style in the home decorator's repertoire. Old wood offers a naturally sympathetic complement to period looks, but its spare, structural quality works equally well in ultra-modern interiors.

When the vogue for stripped floors began, renovating old wood went hand in hand with restoring period features and preserving architectural character: an antidote to all those unthinking and unsympathetic conversions of the post-war period. More recently, such treatment has more or less shed its worthy or historic overtones as home-owners have recognized that a restored wood floor is simply the easiest, and one of the most affordable, means of achieving a pleasant, natural surface that can withstand a fair deal of wear and tear.

The period credentials of bare floorboards do not always stand up to close inspection. Eighteenth-century interiors often displayed simple planked flooring but without the modern array of seals and varnishes, the colour and overall effect would have been a little different. In the Victorian era, when many of today's stripped, bleached or waxed boards were first laid, most wooden flooring in the main rooms would have been fairly well covered by rugs and area carpeting; any exposed wood was generally painted a dark colour to blend in with the background. Stripped floorboards, however authentic they may seem, reveal more accurately a modern desire for simplicity and a renewed interest in natural materials; one day they may seem as indelibly late 20th century as the antimacassar is High Victorian.

STYLES D'EUROPE
LA PORCELAINE EUROPÉENNE
LE XIXe SIÈCLE FRANÇAIS
PHOTOGRAPHS 1931-1955
Oliver Messel
Johannes Vermeer

Above: Stripped Columbian pine creates a warm and uncluttered base in a contemporary living area, and is a good partner for the irreverent leopard-print carpeting on the stairway.

Renovating old wood, by and large, is not expensive – a powerful recommendation for many new home-owners with a sense of style and limited budgets. Stripping old boards that have been hidden from view under a layer of old carpet or lino must rank as one of the most common decorating projects for those in proud possession of an older property. But do not underestimate the amount of the effort involved. Sanding, stripping and finishing original timber floors demands a certain degree of expertise and plenty of muscle. If in doubt, it is always best to bring in the professionals to do the work for you.

The type of original timber flooring most commonly encountered in houses built in the 19th and early 20th centuries are pine boards, laid horizontally across rooms so that they run at right angles to the timber joists. Pine, which is a softwood, has long been a standard constructional material. Softwoods, by their very nature, sand more readily than hardwoods. In older, especially rural, properties, floors may be made of a hardwood, such as oak, often in wide planks. Original parquet flooring is also likely to be made of solid hardwood and can be sanded and refinished in the same way as boards.

Restoring old wood

Once the floor has been given a good overhaul, repaired, prepared, sanded and stripped of previous finishes, there are numerous options available for the final finish, ranging from those which preserve the grainy appearance and tone of the wood, to painted effects of various kinds which cover up the wood's character but widen the scope for colour and pattern.

Above: Original softwood floorboards are relatively easy to restore. Old layers of paint or varnish can be sanded away and the wood stained, bleached, painted, waxed or simply sealed.

Above centre: Stripped pine works equally well with traditional decor.

Above right: Original oak floorboards provide depth of character.

Below left: Kitchens see a good deal of traffic but wooden floors can take a fair degree of wear. These stripped boards are a counterpoint to the gleaming metal fittings.

Below right: Reclaimed pine block flooring makes a heavy-duty surface for a modern island kitchen. Blocks have more utility character than either boards or woodstrip.

Whether an old wood floor has been hidden under other coverings or subject to years of traffic and wear, your first task is to assess its condition and carry out all necessary repairs. This could mean treating rot or structural failure, or simply securing loose boards or filling gaps (see Remedial Work, pages 164–65). Nails which have worked their way to the surface need to be sunk, and the wood cleaned of any old wax, polish or paint. Paint splashes can be removed with chemical stripper and a scraper or sandpaper. If the floor is already painted, you will also have to strip off as much paint as possible before sanding to avoid clogging up the sandpaper. Chemical strippers are quite expensive, and need to be handled with care, but they are the best solution for tackling large or fiddly areas.

At this stage you will be able to assess whether or not the floor requires machine sanding. In rare cases, if the wood is even and has been well protected or if you intend to paint it anyway, hand sanding may be all that is required, working in the direction of the grain.

Top: Old floorboards in a converted loft have been given a new lease of life, sanded to even out irregularities, stained a dark colour and highly polished to make a gleaming reflective surface.

Above: Wood block flooring of mixed woods, with cherrywood inlay used to create an intricate border, lends architectural distinction that marries well with the rich decorating colours.

Left: Stripped boards can be polished in a variety of finishes, from high gloss to a softer, matt look. For a more natural effect, floorboards can also be sealed by applying layers of wax.

Above: Old parquet has a welcome patina and character. Specialist outlets can supply original parquet salvaged from old houses. Slight unevenness of surface and spacing is all part of the charm.

Herringbone parquet must be sanded in both directions, following the pattern. In most cases, renovating old wooden floors will require machine sanding to deep clean, to remove all vestiges of previous finishes and to provide a level surface. Sanding removes the top layer of wood and the poorer the condition of the floor, either in terms of staining or unevenness, the more you will have to sand away to achieve a good surface. For this reason, sanding is best carried out on solid wood floors. Veneered wood may not be thick enough to withstand such treatment.

Sanding is a job for the strong and fit. If you have any qualms about tackling the work, hire a specialist to do it for you: it will add to the cost of the floor, but the overall expense will still be less than most new floor coverings and you will avoid the risk of either injuring yourself or irreparably damaging the floor. Alternatively, you could ask a friend who has successfully sanded a floor to help you out. If you are confident, the basic techniques of using a machine sander are illustrated on pages 168–69.

Decorative treatments and finishes

Most people who go to the bother of renovating old wood do so because they appreciate the inherent qualities of the raw material – the pattern of grain and warmth of tone. Others see a timber floor as a blank canvas for more decorative flights of fancy. Some want a bit of both worlds. All tastes can be catered for by adopting different finishing treatments.

If the 'woodiness' of the floor is what appeals, you will want to preserve those qualities, which generally means doing as little as possible to it. For a seamless look, you can fill gaps between planks with a mixture of wood glue or filler and sawdust, which will blend in with the basic tone and remain flexible enough to tolerate some movement. The floor will then simply require sealing to provide a water-resistant protective coat and make for easier maintenance.

Any other finishing treatment needs to be carried out before the wood is sealed. Painted and decorated floors generally require up to five coats of protective varnish, which means that the area in question will be out of commission for a considerable period.

Bleaching and tinting One of the most straightforward ways of altering the appearance of wooden floors is by bleaching or tinting to lighten or darken the basic tone. It is worth doing a little experimentation to judge the final effect.

Lightening the colour of wood can transform an old floor into an elegant and pristine surface, more at ease with contemporary furnishings. Many old floors are made of pine, which has a warm honey colour when sealed. In certain circumstances, however, sealed pine can look a little glaring and brash, an effect that certain types of seal exacerbate, particularly those containing polyurethane, which tends to yellow with time. Bleaching knocks back the warmth of pine, even to the extent that it begins to resemble a more classy hardwood.

There are a number of different ways of lightening wood. Bleaching is the most extreme, as it takes almost all the colour out of the wood. Another product that gives a similar effect is water-based white pigmented varnish, but it is very expensive.

Alternatively, wood can be lightened significantly by liming. This effect is achieved by rubbing white into the grain of the wood. Any kind of white paint, proprietary liming wax or gesso can be used. Dilute white paint is best for pine. Oak is traditionally limed by first darkening the wood with stain, pigment or ammonia; the liming sinks into the grain to give a weathered appearance. To achieve a pale limed oak, the wood can be bleached first.

In some situations, you may wish to deepen the natural tone of the wood. There are a number of tinted varnishes, wood dyes and stains on the market in a choice of wood colours, from oak to mahogany, which give the wood a more seasoned look. These products can also be useful in spot applications, where a few original boards have been replaced with new timber. Normally, without such treatment, the new patches of flooring will stand out significantly from the rest. To distress new boards, apply a darker stain and then bleach slightly and stain again.

Above left: Whitewash and varnish have been used to lighten this staircase. Work into the grain with a wire brush before applying paint, liming wax or gesso to ensure an even penetration.

Above centre: Wood can be lightened by liming or by applying bleach. When using bleach, it is important to work into the grain and along the boards, so that any discrepancies appear natural.

Above right: Ordinary floorboards can look stylish when stained a dark colour and sealed. Apply the stain to completely clean boards. For the darkest shades, you may need to apply several coats.

Right: Pine has a rather orangey tone when simply sanded and sealed, and the effect can be quite harsh. For subtler results, stain in another wood tone or colour and finish with a matt varnish.

Above and right: A variety of different paints can be used on wood, some more hard-wearing than others. These examples show boards painted in white gloss. Bear in mind that no painted surface is immune to wear.

Colour

The comprehensive selection of wood stains, dyes and paints on the market enables you to make a complete departure from the natural wood palette without losing any of the other virtues of the material itself.

Stain The real advantage of stain is that it soaks into the wood but leaves the pattern of grain still in evidence. The trick is to make sure you work evenly. Stains are available in a variety of forms; oil-, water- or spirit-based. For the best results and the truest colours, it is best to bleach the wood first so that you are starting with as light a background as possible. All stained floors must be sealed (see page 102) to make sure accidental spills do not soak into the wood, showing as unsightly marks.

Paint is the answer if you want a solid, opaque colour. Paintbox primary colours make a splash in children's rooms and playrooms but painted floors can work equally well in different contexts beyond the 'cheap and cheerful' look. Pure white can be soothing and sophisticated in an airy bedroom; black makes a dramatically graphic foil for rugs. Changing the look of an entire room is straightforward enough, simply by painting over a different colour or by introducing another.

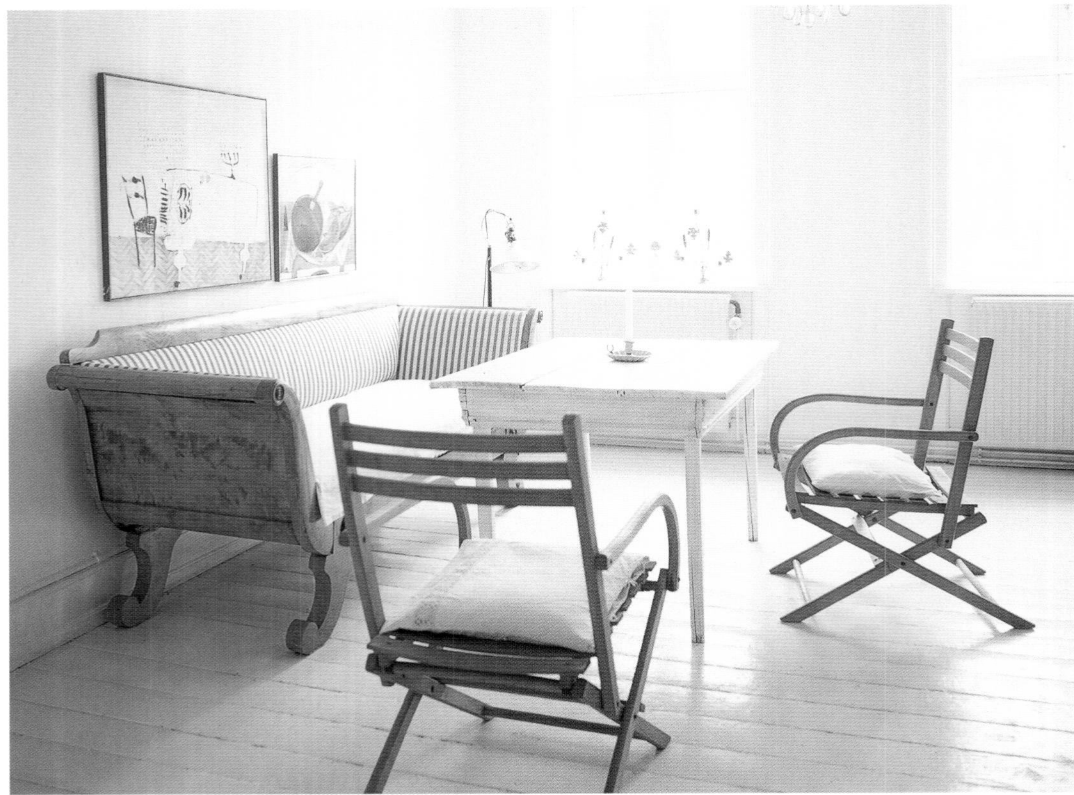

Left: Floorboards painted white in a soft matt finish make the most of natural light and complement the pure and restrained style of decoration. The effect is airy, summery and expansive.

Below, left and right: Wood lends itself to the creation of different decorative effects. These simple but striking examples involve nothing more complicated than painting alternate squares. However, for a crisp look, straight edges are essential.

Although a range of paints can be used on a wooden floor, not all paint is highly wear resistant, so it is best to restrict painted finishes to areas which are not subjected to a great deal of traffic. The final paint finish can be either oil-based gloss or eggshell (which has more of a sheen). Choose carefully: gloss has a shiny, reflective finish which can look chic or somewhat tacky, depending on the context. Eggshell gives a softer, more subtle effect. No further sealing is needed if you use gloss paint, but eggshell needs a coat of matt seal for protection.

There are special floor paints available, which contain epoxy resin or polyurethane for an extra hard finish, although the range of colours may not be so broad as for gloss or eggshell. Yacht or deck paint makes an incredibly robust surface, but it is expensive, hard to apply and takes longer to dry.

Pattern If the sight of bare sanded boards arouses your decorative urges, there are potentially limitless ways of creating different patterns and effects. Many of the paint techniques that can be used to enliven walls will work equally well on floors – especially the bolder ones, from simple geometric patterning to fully realized trompe l'oeil. These effects can be used to add an element of richness and character or to provide a slightly tongue-in-cheek stand-in for a more expensive and luxurious surface. Some are work for the professional; others are reasonably simple to achieve.

Broken colour techniques can be used to give depth and texture to a painted floor. The most successful are bold and forthright, such as spattering and combing; any discreet or subtle effect will be more or less lost underfoot. Stains or dyes are not suitable for these distressed or broken finishes; you need to use paint.

Spattering, as the term suggests, consists of flecking white spirit over still-wet colour, which produces a pebble-like finish. The best effect is achieved by using two or more brightly contrasting shades, slapping down one colour at a time, then scrabbling over the colours before they dry with a sponge to mix them roughly together.

Right: Paint effects do not always have to be precise; the hand-drawn look can be effective, too. This slightly distressed painted effect has its own battered charm.

Far right: Wood stain was used to create this geometric pattern on limed boards. Unlike paint, stain allows the grain to show through, which makes for a livelier surface. The muted colours add interest without being overly dominant.

Combing is similar to dragging and graining, but is more obvious and dramatic looking. The basic technique involves distressing a wet top coat with a comb to reveal patches of a different underlying colour or hue. The result is a pattern of fine lines which gives a sense of movement and texture to the surface. Simple patterns often work best – either stripes or squares (rough accuracy rather than total precision is all part of the appeal). The choice of colours is very important. On the whole, combing is most successful when it combines different tones of the same basic shade. Spattering and combing techniques are illustrated on page 173.

Do not underestimate the time it takes to create a painted floor. You usually need to apply two base coats to undercoated wood and to allow drying time between each coat. When the paint is fully dry, several coats of polyurethane will give a very durable finish.

For painted floor patterns, the regular arrangement of the floorboards lends itself to creating all-over geometric designs reminiscent of tiling. A chequerboard design can be simply executed, using the board widths as a basic dimensional template: one for small squares, two for medium size and three for the largest scale. Paint or stain can be used to build up the design: paint gives a bold, crisp effect, while stain is more textural and muted. Good colour contrasts include the classic black and white, ochre and terracotta or terracotta and black.

Equally effective are patterns which echo the classic slab and keystone arrangement of traditional stone floors. Again, the basic geometry of the boards makes setting up the design much more straightforward. For more complex painted patterns, it is advisable to plan the design first on a scale plan of the floor drawn on graph paper (see page 170).

Borders

A painted border can also lift a plain floor out of the ordinary – and is a cheap and simple method of giving a bespoke finish to a wood floor. Setting in a border (in any flooring material) also provides a useful means of drawing the attention away from less than perfect proportions or irregular features. The trick is to create a regular shape within the perimeter of the floor rather than slavishly follow the precise contours of the room. A border makes the floor look neat and finished and adds just sufficient colour contrast or pattern without being too fussy or insistent. It can also be used to define an area, such as the dining part of an open-plan or through-room.

Stencilling

Stencilling is an excellent decorating technique for floors, either in the form of borders, as edging for stairs or as an all-over pattern. The secret is to opt for crisp, definite designs, scaled to suit the context. Small detailed patterns, especially as borders, can look half-hearted and rather ineffectual. Simple motifs, such as leaves, shells, fleur-de-lys, Greek key and stars have greater impact. There are many pre-cut stencil designs on the market, or you can make your own using stencil card. Play up the contrast by stencilling in paint on a painted floor, or opt for a subtler approach using stain on a stained ground or on natural wood. It may be wise to begin by stencilling a border before embarking on an entire floor.

Wood inlays

More sophisticated decorative effects can be achieved using wood inlays, or by simulating such designs with paint. A feature of wood is that colour, pattern and grain vary so much from species to species. Inlay, marquetry and parquetry are related decorative techniques that exploit the contrasting shades of different woods to create a pattern. In marquetry that pattern is usually representational; parquetry is a term for geometric designs. For inlay, different woods or other materials are inset into solid wood. Marquetry and parquetry, in constrast, involve creating a design using thin pieces of shaped wood and then applying it whole as a veneer over the base. These types of decoration originated as a means of embellishing furniture and were popular in the 17th and 18th centuries. Many flooring companies produce wood inlays for use as borders or medallions. Like any decorative flourish, these are best employed where they will be properly appreciated. Many inlay designs are traditional, but some companies do produce more modern versions.

Above: An inlay-style design simulated in wood stain.

Left: This chequerboard pattern painted onto a wooden floor has been enlivened with stencilled star motifs for additional effect.

Right: Decorative effects need to be displayed where they will be most appreciated. This marquetry-style pattern turns a landing into a focus of attention.

Sealing wood

Sealing is necessary for all new, sanded or decorated wood floors, with the exception of those that are pre-treated or painted with gloss paint, floor or deck paint. Sealing protects the timber from water penetration and makes it resistant to dirt and chemical attack. Test out the seal before you apply it, because many seals alter the colour of the wood and the effect varies according to the seal, the type of wood, and also the light (whether predominantly natural or artificial). In addition, seals are available in a number of finishes, ranging from matt to high gloss, which will also affect the appearance of the finished floor.

Many types of hardwood flooring are supplied ready-sealed and these require no further treatment and only minimal after-care. Untreated wood may need to be sanded once it is laid, then finished with wax, polish or oleoresinous seal. Polyurethane varnish should not be applied until the wood has had a chance to acclimatize to the room where it is being laid; a period of up to twelve months is recommended, but rarely practical.

Some seals are hard-wearing and long-lasting; others require more frequent renewal. Many proprietary seals contain polyurethane, which can cause allergies, and skin and eye irritations when the seal is being applied. Do wear gloves and eye protection when applying a seal. All seals are highly inflammable. If you are concerned about the use of these chemicals in the home, you might consider traditional or natural alternatives based on resin and oil, or wax, although they do not offer such durability.

When applying solvent-based seals, always keep windows open and the room well aired. It is also a good idea to wear a full face mask, which will protect your eyes, nose and throat. Some seals have a long curing time – up to 48 hours – before the floor is ready to be walked on. The number of coats will depend on the product and whether or not the floor has been painted or decorated. You also need to consider how heavily the floor is used. Kitchen floors, which usually take most punishment of all, will need extra sealing, while up to five coats may be needed to protect a painted floor, three for a stained one. Sand lightly with fine-grade sandpaper between coats for the best results. Apply the seal following the direction of the grain of the wood, using a wide natural bristle brush and even strokes.

Types of seal include:

Acrylic This is a water-soluble varnish which is quick drying and non-toxic. It is also easy to apply. It needs to be waxed two to four times a year.

Alkyd resin A non-yellowing, non-toxic and quick-drying seal. Like acrylic, it needs to be waxed two to four times a year.

Button polish (shellac) The traditional floor seal, but it becomes brittle and is easily scratched and stained.

Epoxy resin This seal is rather yellow in tone and slow to dry but it is very hard-wearing.

Natural sealants These are safe, environmentally friendly products composed of resin and oil. Some include wax. They are water-resistant and can be tinted with natural stains or colours.

Below left: Pine tends to look harshly orange when sealed with polyurethane. Bleaching the wood first can offset the tendency.

Below: Try out different seals on offcuts and judge the effect in the room you intend to seal, before tackling the whole floor.

Below, right and right: Any wood floors which are likely to see spills of water, such as a conservatory or kitchen, must be treated with several coats of seal. Heavily used areas will likewise need several coats – and may still need renewing fairly regularly.

Oleoresinous These varnishes are a blend of tung oil and resin (usually phenolic resin). They are clear but with a slight yellowish tinge and are slow-drying but simple to apply. They are not the most hard-wearing of seals for wood but they are easy to patch.
Polyurethane This is one of the most common types of seal. It brings out the yellow tones of wood and may also yellow with age. Polyurethane is hard-wearing but it can cause irritation when applied. It is easy to patch.
Urea formaldehyde These transparent lacquers are ideal for light-coloured or bleached woods. They are hard-wearing, but difficult to patch-repair.
Wax Special natural floor wax, consisting of a mixture of bees- and plant waxes. It is durable, water-resistant and has a pleasant scent and anti-static properties. Liquid beeswax is also available.

Maintaining wood

Properly sealed and finished wood is easy to keep clean on a week-by-week basis. Vacuum or sweep up loose dirt regularly and wipe the floor with a damp cloth or a mop. Avoid overwetting the floor and mop up spills as they occur. Use a mild detergent on stubborn marks if necessary.

If you have a waxed finish, the wax polish will need to be renewed at intervals – say every couple of months. Most seals will last far longer. Areas of heavy traffic may wear out faster, but certain seals, such as polyurethane, can be patch-repaired with ease. Use cups under the feet of furniture, and do not drag heavy objects over the floor. Avoid stiletto heels at all costs – even at the risk of embarrassing your guests!

Sheet & Soft Tiling

BOSCH

Sheet and its close associate soft tiling are the great cover-ups of the flooring world. This is not to imply that they hide a multitude of sins — no floor treatment will look or perform its best over a shoddy base — but they do provide a simple, quick and relatively cheap way of creating a seamless or integrated effect while delivering maximum convenience.

Often, the distinction between tiling and sheet flooring is more practical than visual: tiles can be abutted without obvious joins while sheet flooring can be patterned to resemble tiles. Many different materials are now available in these two basic formats. There are natural or near-natural types, such as linoleum, cork and some forms of rubber. Others, vinyl for instance, are just about as synthetic as you can get. The cost varies considerably. Linoleum, rubber and top-quality vinyl are at the upper end of the scale and can be as expensive as the best carpet; cork is significantly cheaper. Quality also affects performance: linoleum lasts for years; cheap vinyl has a short life. Installation, too, varies across the range: laying soft tiles is the type of job any reasonably competent person could tackle in an afternoon (see pages 176–77), while putting down heavier sheet flooring over a large area is skilled work for the professional.

All these forms of tile and sheet have certain common attributes. They tend to be soft, warm and comfortable underfoot, with the cushioned varieties providing a high degree of give. They are also lightweight, so you don't need to worry about the load-bearing capacity of a particular floor.

By and large, most of these materials lack the intense beauty that is the hallmark of more authentic types of flooring such as stone and solid wood – although some people get distinctly emotional about linoleum. (Leather, of course, is in a class of its own.) Attractive and appropriate rather than irresistible, sheet and soft tiling score highly for ease of maintenance and sheer practicality. With the exception of cork, these materials come in a staggering range of colours, patterns and textures. Sheet linoleum, in particular, offers great scope for custom designs. The mainstream market, however, tends to be dominated by designs which simulate other more innately soulful materials, with countless variations on the theme of wood-, stone-, marble-, quarry- and brick-effect.

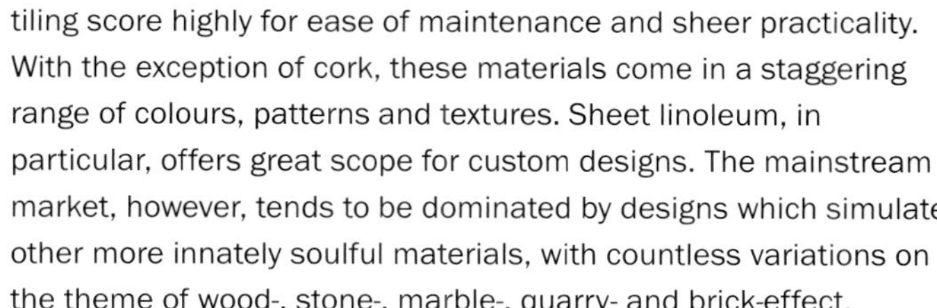

The whole issue of simulated flooring has a tendency to bring out the style council in full force. The less convincing simulations do have a rather apologetic, cringing quality that many people find depressing. Yet at the same time, really skilful vinyl copies (which can be quite expensive) have their own built-in letdown. Ultimately, no matter how visually close such simulations are to reality, the fact that they can never sound, smell or feel like the real thing will inevitably give the game away. With time, this perception will only become more acute. Synthetic flooring does not wear like its natural counterpart: its pristine perfection when new lacks the random depth of character of materials such as wood and stone and once it begins to degrade it tends to lose charm completely.

That said, some simulations do eventually acquire an odd sort of 'rightness' just through prevalence, while others may be prized by devotees of kitsch. In terms of style, though, it can often be a better idea to choose a synthetic floor with a design that displays an abstraction of pattern or which summarizes the qualities of another material rather than one that is outright pastiche. Marbleized tiles, for example, have the cloudy veined pattern of the stone in graphic form without attempting to provide a complete replica of it. There are also plenty of exciting colours, patterns and finishes that are not trying to be something else and are all the better for it. It is, however, unrealistic to imagine you can deceive anyone for long into believing that they are walking across oak parquet when they are in fact treading on a clever vinyl lookalike. But for those who wish to take the appearance of a natural floor into areas such as kitchens where maintenance would otherwise be prohibitively demanding, a good simulation may do the trick.

Although this family of flooring materials may essentially comprise workhorses rather than thoroughbreds, style-council dictats or no, there will always be room for them somewhere in the home at some stage in life. Undoubtedly, many of these materials are chosen as a stop-gap or as a sensible solution for hard-working utility areas, but there is no reason why they should not be viewed more positively. Where they are employed with confidence and flair, the style dividends can be as high as the practical benefits.

Far left: Lino is a practical, wholly natural material with excellent environmental credentials.

Left: Vinyl, lino's synthetic alternative, comes in a wide range of colours, patterns and textures.

Cork

Soft, quiet and comfortable, cork makes a good-looking practical floor in many areas of the home. Derived from the bark of the cork oak (*Quercus suber*), an evergreen native to Mediterranean areas, its warm neutral tones and open texture have a sympathetically natural appearance. Cork oak is grown commercially for the purpose of extracting the bark – a process that does no harm to the tree, which simply regenerates it. The cork bark is then granulated, pressed with binders such as resins, and baked.

Anyone who has ever pulled a cork from a bottle and seen it spring back into shape can appreciate the remarkable resilience of this material. It is this quality that makes cork extremely comfortable to stand on and walk across, reasonably resistant to indentation from heels or furniture legs, and quiet too – valuable attributes for use in children's rooms, kitchens, bathrooms, cloakrooms and hallways. In all of these respects cork scores more highly than lino. It is also anti-static and non-slip, even when it is wet or polished.

Cork can be used unsealed in the bathroom, where the effect is similar to having a large absorbent bathmat permanently in situ, but normally it does require sealing to prevent dirt becoming ingrained. In areas of high traffic, such as around the kitchen sink or by the front door, cork can look shabby rather quickly and for this reason it has acquired a rather drab image in some people's minds. However, with appropriate use and treatment, a cork floor can maintain its handsome and unpretentious appearance.

Most cork is sold in the natural shades of pale honey to dark brown, but there are coloured versions including charcoal, green and blue. Using contrasting colours or an inset border will add an extra style element. Tiles (sealed or unsealed) are the standard format and the thicker the tile the more hard-wearing the floor will be. Cork wall tiles are not suitable; the cork must be flooring grade. Cork sheet or 'carpet' with a jute canvas backing is also available but it is more unwieldy to install. Some manufacturers produce composite cork flooring which offers much greater wear-resistance and durability, but at the expense of a slight compromise of the natural qualities of cork. Such flooring consists of a layer of compressed cork sandwiched between a vinyl backing and a cork finish or veneer, sealed with a layer of clear vinyl. It is available in tile or 'plank' form and in a broader range of colours. There is also veneered tongued-and-grooved strip flooring, with cork substituting for wood as the final finish over compressed cork, MDF and a plywood base.

Laying

Natural cork tiles are easily within the scope of competent amateurs. They are lightweight and easy to cut. They require a dry, level sub-floor, such as floorboards covered with a layer of hardboard or plywood. Paper felt laid over the hardboard will prevent nail heads working through. Concrete sub-floors at ground level will need a damp-proof membrane. Cork cannot be laid over underfloor heating, which would cause the tiles to lift. To allow the material to adjust to the ambient temperature, cork tiles need to be stored in the room in which they will be laid for 48 hours before gluing with the recommended adhesive. 'Planks' should also be glued. No expansion gaps are necessary. Organic adhesives can be used. Tiles may need to be weighted until the adhesive has set.

Cork-veneered strip flooring can be laid as a floating floor, without nailing or gluing the strips to the underlying surface. The advantage is that the strip floor can be taken up at a later date without affecting the existing floor beneath. The tongued-and-grooved planks are slotted together, with a PVA adhesive used along the tongue to seal the floor against dust. An expansion gap of 5–10 mm is needed around the perimeter of the room.

Sealing and maintenance

Sealing is strongly recommended to protect the cork from soiling and to make it water-resistant. After laying, allow about 48 hours before sealing to give the adhesive time to cure. Three or four coats of polyurethane or polymer sealant will prevent the surface from degrading: again, you need to allow sufficient time between each coat. Presealed cork can be given a single extra coat of seal after laying to prevent moisture from penetrating joints. Alternatively, unsealed cork can be wax-polished or dressed with organic primers. Cork flooring with a vinyl finish requires no subsequent treatment.

Despite being cheap and straightforward to lay, a cork floor demands dedicated upkeep to preserve its appearance. It is important to make sure the cork is kept clean and free from grit, which might break down the seal and allow dirt and moisture through. Sweep or vacuum the floor regularly, wipe with a damp cloth, and polish occasionally and sparingly.

Above and left: Cork has seen something of a revival in recent years, after a few decades of obscurity. Well sealed, it is both handsome and practical. Other advantages include the fact that it makes a comfortable, warm surface underfoot and has excellent properties of sound insulation. It's also a natural product, which makes it a sound choice for those with ecological concerns: the basic ingredient is harvested from the bark of the cork oak, a process which does no harm to the tree. Aesthetically speaking, cork suits contemporary styles of decoration, particularly when teamed with area rugs.

Rubber

With its tough, industrial aesthetic, rubber flooring has long been a favourite with modern architects and designers. Widely used in stations, airports and other public transit areas (even on spacecraft), studded or embossed rubber migrated to the home with the arrival of hi-tech in the late 1970s. Hi-tech, as a trend, did not last for long, but rubber flooring is once again showing signs of a revival among those at the cutting edge of interior fashion, mainly because a wider range of colours is now available.

You don't need to be in the stylistic vanguard to appreciate the considerable practical advantages of rubber as a domestic flooring. Hard-wearing, water-resistant, burn-resistant, extremely resilient, quiet and warm, it has all the qualities anyone would wish from a utility floor. Naturally, it is in bathrooms, kitchens and other service areas where its use is most prevalent. Nevertheless, there is plenty of potential for a wider application.

Natural rubber is less consistent and harder to colour than synthetic varieties and most of the rubber now sold for use in flooring applications contains little if any natural elements. Instead it consists of vulcanized synthetic rubber, pigment, silica and china clay. Some types also come with fabric or foam rubber backing. Rubber is most commonly available in tile form, but wide rolls, stair

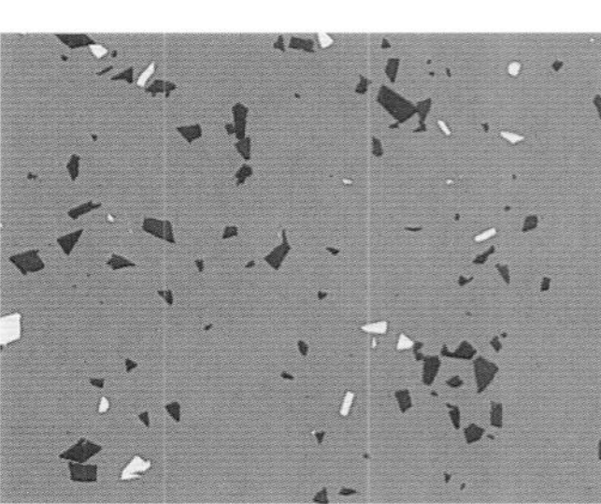

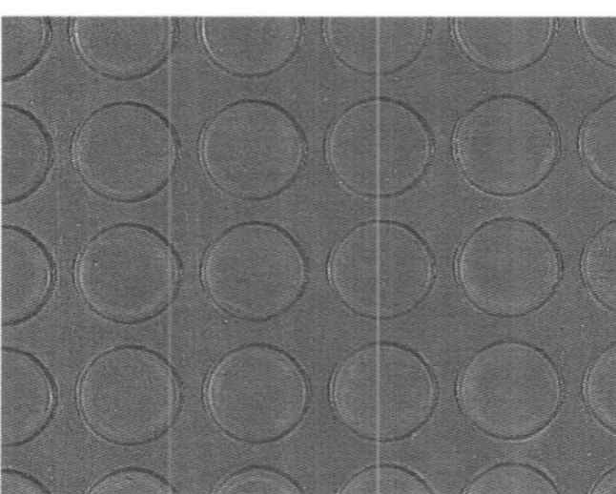

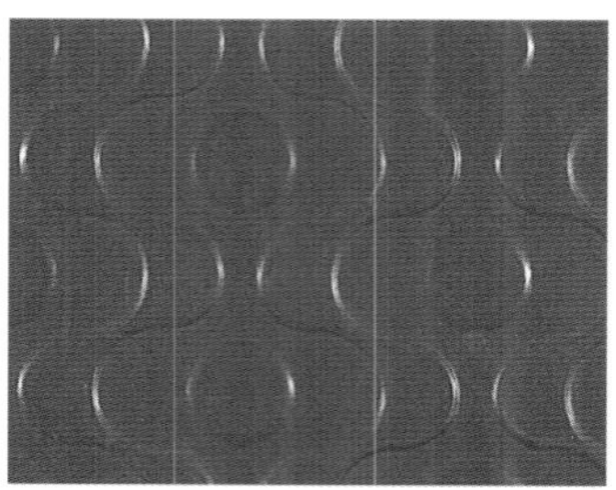

Right (details): Svelte and sophisticated, rubber flooring has shed some of its harder industrial aesthetic to be reborn in a wide range of brilliant colours, abstract patterns and textural relief patterns. The textured versions add grip, and are recommended for stairs or areas which are likely to become wet, although the relief patterns tend to trap dirt and require more regular maintenance than the smooth finishes.

Left: This jolt of lemon yellow makes a bold modern statement in a living area.

Centre: Rich brown rubber flooring complements the cool blue-green kitchen decor. Rubber does stain, so extra care must be taken where spills are likely and sealing is advisable.

Far right, above and below: Ideal for bathrooms with attitude, rubber flooring provides a depth of pure colour. Rubber can become slippery when wet, however.

tiles and runners are also produced. It varies in thickness, which affects the price. The best grades will cost you as much as top-quality vinyl or wood-block flooring.

In the first flush of enthusiasm for rubber flooring, black and grey were popular colours, accentuating the severe hi-tech look, with the occasional brilliant blue, grass green and scorching yellow making a more vivid statement. Nowadays, rubber flooring comes in a vast range of colours, from bright paintbox primaries to soft, subtle tones that lack any hint of the factory floor. There is an equivalent variety of finish and pattern. As well as smooth, matt solid colours, there are various raised textures including round studs, ribs, grooves and designs reminiscent of metal treadplate which are intended to improve grip, and broken colour effects such as marbling, striation and flecking designed to disguise dirt.

Studded rubber is a good choice for any area likely to become wet — smooth finishes can become very slippery. Dirt and food spills do have tendency to accumulate around relief patterns, however, and extra maintenance may be required to keep the floor clean. Unpolished matt rubber marks easily. There is a special grade available for laying over underfloor heating, but the warmth may bring out a rubbery smell.

Rubber flooring is often laid as a single-colour floor, but tiles can be bought in contrasting shades to make a chequerboard effect or in a more random medley of colours. Sheet rubber lends itself to the same sort of cut patterns as sheet lino.

Laying

Rubber is supple, easy to cut and shape, and lasts forever. It must be laid over a dry, flat sub-floor — either floorboards covered with hardboard or plywood, or concrete. Solid floors at ground level will need a damp-proof membrane. Both tiles and sheet need to be bonded to the sub-floor with adhesive. Recommended rubber adhesives are usually applied both to the sub-floor and the flooring, and the two surfaces brought together after a specified period. After laying, allow about 48 hours before polishing to give the adhesive time to cure.

Maintenance

Tiles may come with a surface layer of dust which has to be removed with a special cleaner. Rubber can be left matt for a warm, soft look or polished with water-soluble wax emulsion. Natural rubber is readily marked by fats and solvents, but even synthetic rubber can be damaged if left unpolished. Use polish sparingly and renew it from time to time.

Left: Rubber comes in a range of textures. Studded rubber, as here, is a good choice for areas that may get wet; smooth rubber can be slippery.

Below: Rubber can be polished with a water-soluble wax emulsion. Waxing intensifies its colour and increases stain-resistance, which is important in kitchens and dining areas.

Bottom: The usual purity and simplicity often associated with bathroom decor has been cheerfully ignored in this chequerboard design of coloured rubber floor tiles.

Linoleum

Linoleum is the Cinderella of flooring materials. In the not-too-distant past, it was fairly universally reviled as the grim, cracked utility covering of old hospital corridors, but in recent years, manufacturers have virtually reinvented the product, transforming a brittle and often lurid material into a sleek robust flooring available in a huge range of colours and patterns. This reincarnation has coincided with increased public awareness of environmental issues. Linoleum is a wholly natural product which just happens to deliver many of the same practical advantages of similar synthetic types of flooring, along with other benefits uniquely its own.

Linoleum has had a respectable history since its invention in 1863 by the Englishman Frederick Walton. It grew in popularity from the latter decades of the 19th century onwards, when it was first put into mass production, serving as a cheap, practical and more durable alternative to oilcloth. It proved exceptionally hard-wearing; examples of old lino floors dating back to the early decades of this century are still discovered from time to time in out of the way church halls and working men's clubs. Lino was often printed in designs that reflected the style of the period; Bauhaus architects were fascinated by the material and some, including Mies van der Rohe, Behrens and Hoffman, produced designs for it.

Earlier forms of lino were, however, brittle in comparison to today's versions and dreary cracked tiles or sheet curling at the edges helped to give the flooring a depressing second-rate image. The image problem was not helped by the misleadingly termed 'printed linoleums', cheap ersatz products of very poor quality which were essentially painted bituminous felt. The introduction of vinyl flooring after the Second World War marked the beginning of the long decline of lino in popularity.

The contemporary revival reflects technical improvements which have greatly increased the scope of colour, pattern and design. At the same time, attention has been focused on the many other estimable qualities of lino. It is not only a natural product, but a healthy one. It is anti-static, which means it does not attract dust and dust mites, which can trigger asthma or allergenic reactions. Equally important, lino is anti-bacterial, naturally killing off germs that come into contact with it, an asset which has long made it a popular choice for hospital flooring. It is warm, quiet and resilient, comfortable to walk on and reasonably non-slip even when wet. All of these qualities are enhanced in the thicker, more expensive grades; in addition, there are special varieties such as cork-backed lino, which give a high degree of cushioning and sound absorption, and hardened grades which are very resistant to indentation and burns. All types of lino can be used over underfloor heating.

Linoleum takes its name from *Oleum lini*, the linseed oil derived from flax. To make lino, linseed oil is oxidized and natural pine resin (rosin) is added as a hardener. The lino 'cement' is then mixed with powdered cork, which provides its insulating qualities and flexibility, and wood flour and powdered limestone, which lend hardness and strength. Pigments are added for colour. The raw material is then pressed onto a woven jute or hessian backing and left for several weeks in a drying chamber where it is baked at high temperatures. It is then ready to be cut into rolls.

Lino has a naturally grainy matt finish, which can be buffed and polished to a glossy sheen. It is available in an enormous range of colours, twice as many in sheet as in tile, as well as a variety of finishes, including marbled, streaked, striated, flecked and mottled. Colours range from intense hues to pale tones but all have a natural, slightly soft quality. New designs include geometric patterns such as plaid, gingham, keystone, and box (or tumbling block) as well as a range of borders. The latest technology has revolutionized the creation of inlaid designs: very complicated patterns and well-defined motifs can be realized by computer-controlled precision cutting.

Lino is damaged by damp; careful laying is important to prevent water penetrating joins and rotting the floor from below. Sheet lino can be hot-seam welded to make it watertight; tiles must be

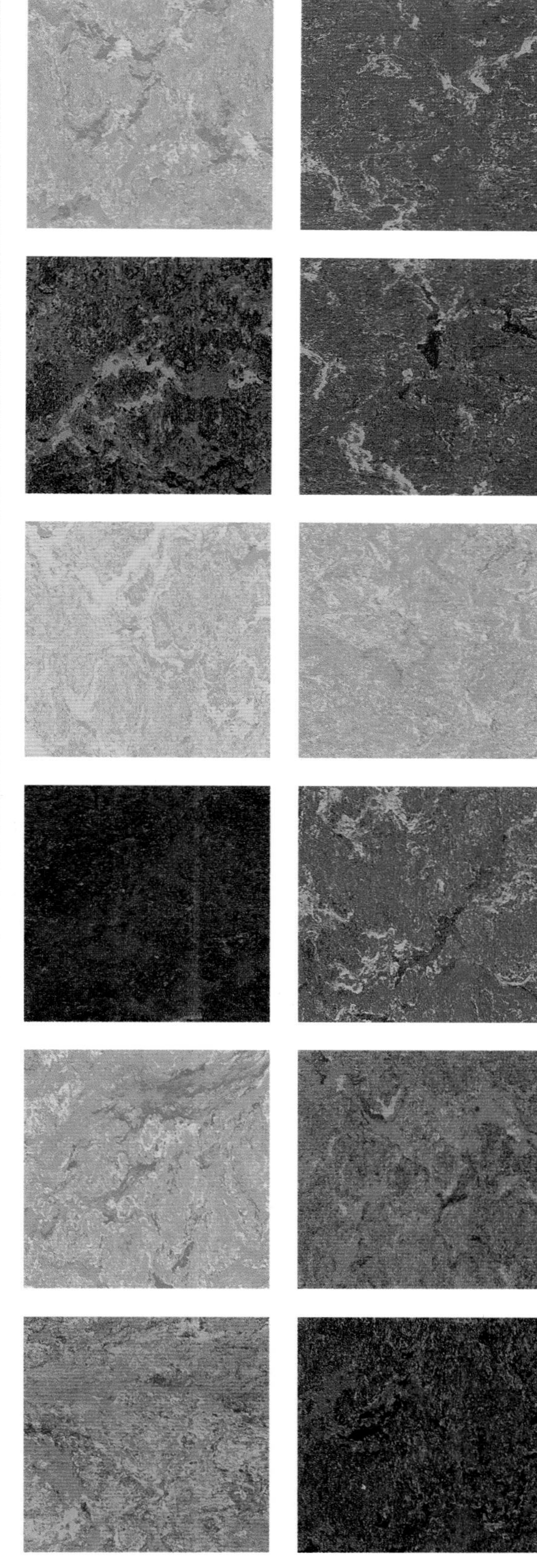

abutted tightly. While the material itself is not exorbitantly expensive, the cost of professional laying can push the final price of a finished floor into the upper bracket, particularly in the case of an inlaid design. Tiles can be laid by most skilled amateurs, but sheet lino is heavy and unwieldy and needs expert attention. Once laid, although it is proof against most household chemicals, lino can nevertheless be damaged by a variety of solvents.

Like most natural products, lino improves with age. The linseed oil used in its manufacture continues to mature for up to ten years, which means that lino actually gets tougher as time goes by.

Left: Borders and central motifs lift linoleum out of the ordinary. Practically speaking, linoleum makes an excellent kitchen floor, being naturally anti-bacterial and easy to maintain.

Above: Plain lino has a sleek, stylish contemporary look.

Right: Linoleum has a pleasing grainy matt texture, usually with a slightly soft quality. It comes in a vast range of colours and finishes.

One particular feature of linoleum is what is known as 'stove yellowing', a temporary discoloration which arises as part of the maturing process of the material in the ovens. It is especially noticeable on white, blue or grey shades, as a yellow tint or 'oxidation film', but it disappears within a few hours of the material being exposed to sunlight. In dark areas, such as basements, the tint may last for up to several weeks. The effect may be particularly apparent when comparing a section of laid floor with unlaid lino from the same batch, but does not reflect discrepancies in the material itself.

Linoleum needs a period of acclimatization before laying, up to 48 hours at room temperature. Tiles should be laid butted up tightly against each other. Seams have a tendency to get tighter with age, rather than to shrink, which helps prevent moisture from getting through. Lino in sheet form is generally available in 2-metre widths. It is heavy and not as pliable as vinyl or rubber, which makes installation a professional job. In sheet or tile form, lino must be glued in place with emulsion, synthetic rubber resin or powder adhesive. Sheet lino should be firmly flattened with a floor roller and any joins hot seam-welded.

Inlaid patterns

Original designs are made by cutting and piecing sections of sheet lino or by inlaying a decorative border. The cut elements are then supplied ready to install by a professional floorlayer: the end results can elevate a lino floor from a position backstage into one that well deserves the limelight. Many manufacturers offer a custom service for the commissioning of individual patterns. The price will vary according to the complexity and detail of the design and the number of colours employed. In the past, creating an inlaid design depended exclusively on the craft skills of the floorlayer. Nowadays, designs are fed into a computer which controls the cutting of the inlaid pattern with high-pressure water jets. Inlaid patterns range from simple curved contours, pathways and random insets to the more complex and detailed designs facilitated by precision cutting.

Installation is a highly skilled operation. The background lino is first dry-laid on the floor and the inlays placed on top and marked accurately in position. Then the background is removed and the inlays are stuck into their measured positions, rolled and weighted if necessary. Finally, the background is cut so that it overlaps the inlay slightly, stuck down in place and trimmed to fit. Any small gaps can be filled and polished off with linoleum dust blended with a dilute emulsion.

Laying

Linoleum should be laid over a dry, flat sub-floor, such as floorboards covered with hardboard or plywood, or concrete. Concrete at ground level needs a damp-proof membrane. The thicker grades of linoleum will make up for slight imperfections in the base, but unevenness needs to be remedied as it may cause the lino to crack.

Left and below: Lino improves with age, becoming harder and more wear-resistant. Sheet lino must be laid by a specialist.

Right: Cheerfully domestic, these large blue-grey and white linoleum tiles complement a retro-style kitchen perfectly.

Caloric

Vinyl

Vinyl is the term given to a range of synthetic flooring products which contain some proportion of polyvinyl chloride (PVC). First developed in the 1950s, PVC is a thermoplastic, in other words a type of plastic which retains the ability to be softened by heat, a characteristic that gives it flexibility. The actual amount of PVC in vinyl tiles and sheet varies; the higher the percentage, the better the performance and quality and the higher the price.

Vinyl is a very popular mainstream flooring. Affordable, easy to install, non-allergenic, simple to maintain and with a reasonable lifespan, it offers a straightforward and practical solution to many flooring needs. The choice of colours, patterns, textures and 'effects' on the market is huge. Vinyl is available in a range of thicknesses and tile sizes, and standard sheet widths of 2, 3 and 4 metres, which makes it possible to cover most floors seamlessly.

Practicality is one of vinyl's major selling points. Vinyl is waterproof, resists oils, fats and most household chemicals, and makes an easy-care, all-purpose flooring. Cushioned vinyl, which has an interlayer of foam, and vinyl backed with foam rubber provide extra resilience and better sound and heat insulation. Vinyl can be used over underfloor heating and sometimes laid directly over existing floor coverings.

Above: These tiles, which consist of a printed design coated with vinyl over a cork backing, come in a wide range of photographic designs, including roses (as shown), grass, leaves and stones. Perhaps a little too intrusive used on a large scale, such patterns are ideal for small, enclosed areas such as bathrooms and loos.

Right: Plain sheet vinyl makes a no-nonsense floor in hard-working areas of the home, such as kitchens and bathrooms.

Far right: Textured vinyl makes a vivid non-slip covering for stairs.

As far as appearance is concerned, simulations of natural materials dominate the quality end of the market. In fact, in the literature of some manufacturers it can be difficult to find any mention of the synthetic origins of the product, with terminology such as 'marble design' or 'beech effect' hinting at more worthy antecedents. Many of these simulations are very highly realized, with shading, graining and textural variations faithfully rendered. Vinyl wood-effect flooring even comes in random 'planks'. Popular simulations include all forms of wood, from parquet to rustic boards, as well as the ever-popular marble, slate, brick, quarry tile and terrazzo. The vinyl versions, unlike the natural materials they mimic, can be readily combined in the same floor, or as borders. Custom design services are available from leading manufacturers, with computer realizations offering a preview of the final effect, as well as an installation plan and a cutting list to make laying simpler.

Such high-level simulations do not come cheap, although in comparison with non-synthetic materials the price is much more favourable once the cost of laying and maintaining the floor has been taken into account. It is important to remember, however, that what may look utterly convincing in a full-colour brochure will inevitably lack some degree of authenticity in real life. Marble-effect vinyl, for example, is warm to the touch; beech-effect vinyl will never acquire more patina than what it has in simulated form already: the beauty is literally only skin deep. To some people the whole concept of simulated materials is an anathema; others relish the opportunity to live with the appearance of a natural floor without the bother of looking after it.

Simulated designs are not the only types available. There are simple geometric patterns; mottled, speckled, metallic and flecked finishes; and lively contemporary styles. Particularly striking are those vinyls which achieve almost a three-dimensional effect by suspending coloured PVC granules or glittering quartz flakes under a clear surface layer — similar to industrial or contract vinyls, which use chips or fragments of natural fillers to increase slip-resistance (smooth vinyl is slippery when wet). Such designs sparkle as they catch the light in different directions; one style even glows in the dark. The whole notion of simulated flooring has been given an original, witty twist by one designer who has come up with the idea of photographic tiles. Protected by a durable vinyl coating and backed by cork, the tiles show shots of outdoor landscapes at close quarters: a pattern of waves; a sandy beach complete with shells; rocks and pebbles; and wildflowers — ideal for creating a watery bathroom floor or rocky entrance.

For anyone who relishes the cheap and cheerful look of plastic, there are also PVC runners and mats. These temporary cover-ups come in jolly colours and 'weaves', as well as in neutral shades and rougher textures, which have the look of natural fibre matting.

Despite its high reputation for practicality, vinyl does have some disadvantages. The quality of the sub-floor matters a great deal: many types of vinyl are not thick enough to absorb minor discrepancies and underlying ridges, bumps or protruding nail heads will show up on the surface where they may cause patches of wear or even holes.

Any sign of wear has an entirely negative effect on the appearance and performance of a vinyl floor. Unlike natural materials, which may mellow with time, worn vinyl merely looks shoddy. For this reason, you must ensure that the right cleaning products are used and that the floor is protected from grit, as well as the type of spills most likely to damage it. Black rubber heel marks can be particularly detrimental; if not tackled immediately, anti-oxidants in the rubber can stain the floor permanently. Rubber-backed rugs and mats can cause similar problems. Vinyl is also badly damaged by cigarette burns.

Finally, vinyl is not a material for anyone who is concerned about environmental issues. Soft plastics such as PVC have been identified as among those most likely to 'offgas', that is, release potentially hazardous chemicals into the atmosphere. If vinyl catches fire, the fumes are toxic. The manufacture of vinyl also consumes non-renewable resources of petroleum and natural gas and the product itself is non-biodegradable.

Laying
See pages 176–77.

Maintenance
Vinyl requires no sealing, but you must follow correct maintenance procedures to prevent the material from becoming worn too quickly. The surface can be torn by dragging heavy items over it and premature damage can be caused by the use of the wrong cleaners. See pages 180–81 for advice on its upkeep.

Left and far left: At the upper end of the price range, high quality vinyl, such as this Amtico, can be very convincing. Both of these examples show a slate-effect vinyl, one 'textured' and the other resembling a matt honed slate.

Leather

Utterly sublime, extravagant and expensive, leather tiles are the height of sophistication. For the home that has everything – and the home-owner with deep pockets – leather makes an unusual and truly classy floor, bursting with warmth and character. It naturally works well in the clubby atmosphere of a study or library, but also with clean, modern lines in contemporary living spaces.

Leather is warm, comfortable, resilient, sound-absorbent and hard-wearing; underfoot it feels and sounds a little like walking across a wooden floor. Like other natural materials, it improves with age and use, acquiring a deep, rich patina with repeated waxing and buffing. It is, however, far from a utility covering and should not be laid in kitchens or bathrooms. The price is in the upper echelons, comparable to that of natural stone or the best-quality ceramic tile.

The tiles are made of steer hide, cut from the centre portion, which has the toughest fibres, cured by traditional vegetable tanning and dyed with aniline dyes. Colours include natural, dark red, rust, brown, dark green and black; and the finish can be either smooth or antique-textured to resemble boar hide. Sizes and shapes of tiles vary, enabling a variety of patterns to be created.

Leather should be laid on a dry, even base, preferably on hardboard or plywood. It is bonded with contact adhesive, with the tiles laid to a line and abutted closely together. Slight variations may occur in tile size and shape and any gaps can be filled with carnuba wax, which will act as a grout.

Waxing provides basic moisture-resistance and spills, if tackled immediately, will wipe off. Scratches are inevitable, but add to the character; a well-used floor has great depth of appeal.

Left and above: For a look of understated luxury, nothing is more effective than leather. Restricting the decor to a warm palette of neutral shades, from honey to biscuit, keeps the effect chic and classy.

Far left: In a Japanese-inspired interior, leather flooring adds a material sensuality to complement the translucent rice-paper screens and wood detailing.

Soft floors

Carpet, rugs and natural fibre coverings occupy the 'soft' end of the flooring spectrum. These materials tend to be warm, comfortable, quiet and enclosing – as well as fairly impractical for hard-working utility areas such as bathrooms and kitchens. In almost every other respect, however, generalizations are largely meaningless: price, composition, performance, colour, pattern and texture vary widely.

Wall-to-wall carpeting is synonymous with the modern lifestyle. Along with central heating and open-plan spatial arrangements, fully carpeted rooms became a signature of a certain kind of post-war interior – informal and above all comfortable. What was once so progressive can, however, seem a little passé. Like other furnishings that were status symbols a few decades ago, carpeting has lost some of its former glamour. Its appeal has never waned for use in bedrooms, but carpet is now less of an instinctive choice for living areas, losing ground in style-conscious circles to wood and natural fibre coverings. In recognition of the shift in popularity, some manufacturers are now producing wool carpet in 'sisal-look' weaves to give a classic product a more updated image.

The runaway flooring success of recent years, natural fibre, encompasses a range of materials from bristly coir to the smooth sophistication of woven jute. With their natural credentials and historic pedigrees, these types of soft flooring satisfy the contemporary desire for authenticity on both counts. Although stronger shades increasingly feature, a pleasing neutrality is the norm, a muted palette which is given depth by the texture of both weaves and fibres, albeit at the expense of a slight compromise in comfort. Manufacturers and suppliers are rapidly extending the choice of colours and patterns, as well as improving practical performance of natural fibre coverings.

Loose-laid rugs are a study in themselves, from the priceless Persian with its tribal iconography to the cheap machine-washable cotton dhurrie. Rugs may seem accessories of the flooring world, offering instant uplift, warmth, colour, comfort and drama, but they are often much more than a mere accent or finishing touch. A beautiful rug can be the centrepiece of the entire room, its colours and design forming the foundation for the decorative scheme.

With the exception of rugs, most soft floors benefit from professional laying. At the cheaper end of the market, foam-backed carpet can be laid by an amateur (see pages 178–79), but most carpeting and natural fibre floorings are as unwieldy as sheet flooring and demand expertise to fit. Also, most types of soft flooring stain easily and, while stain-inhibition treatments are available, you need to be more vigilant to make sure they retain their good looks.

Soft flooring appeals to our sensual nature, inviting us to kick off our shoes and lounge on the floor. A home fitted with carpet from the front door to the attic steps might be a deadened, overly cushioned place (not to mention an impractical one), yet a home with no soft surfaces would be unyielding, rackety and harsh. Textile or woven coverings offer the opportunity to change the pace and the mood.

Left: New custom designs extend the scope of carpet pattern into exciting dimensions. Random or abstract designs marry well with the modern aesthetic. There are now a number of high-quality manufacturers who offer a service to produce carpets and rugs to original specifications.

Right: The work of a Finnish designer, these luminous soft floor coverings are made from paper twine.

Far right: Area rugs provide interest and can be the start of a decorative scheme.

Above: Contemporary rug design is a flourishing field. The work of classic modern designers such as Eileen Gray is still in production, along with many innovative designs by newer talents.

Below: Traditional stair runners complement the country house look. Vibrant colours are set off with pristine white woodwork.

Carpet

Carpet conveys a sense of warmth, intimacy and comfort which many find irresistible – and it is one of the most popular of all flooring options. Although it has become a much more predictable and immediate choice for the average homeowner over the last few decades, it still evokes a feeling of luxuriousness and indulgence. Carpet is the softest flooring material available and positively encourages the most direct contact – whether it is bare feet getting out of bed in the morning, bare young knees crawling around on the floor or simply the easy informality of wandering around the house without shoes.

Far left: Carpet tends to create a sense of enclosure, which is good for unifying spaces.

Left: Soft, warm and comfortable, carpet accentuates the soothing intimacy of an all-white bedroom.

Below: Hard materials can be noisy on stairs and carpet often makes a more practical solution. This elegant herringbone design with contrast banding adds a graphic edge. Stair carpeting can also be loose-laid, secured firmly in place with stair rods, which means the carpet can be turned periodically to even out wear.

Right: The textural quality of this weave adds character and visual interest when colours are muted and natural.

Far right: A specially commissioned carpet in pale oatmeal cord makes a suitably discreet partner for the light, airy decorations.

Carpet did not become a term applied to floor coverings until the mid-18th century, when the first carpet factories were established in continental Europe and Britain, making widths that could be pieced or sewn together to cover an entire floor. Early English manufactures included Wilton, in Wiltshire, founded around 1740, which made carpet with a cut-pile surface, while other factories, notably at Kidderminster in Worcestershire, made so-called 'in-grain' or 'Scotch carpet', a flatweave variety with no pile. Carpet – in the sense of a large unfitted covering with a pile surface woven in a single piece – also began to be made in the West at this time, notably in France at the Savonnerie factory near Paris and in south-west England, at the Axminster factory. These carpets were luxurious and expensive, much more costly than the Oriental or Turkish versions. (For the purposes of this chapter, the term 'carpet' is used in the modern sense to refer to an integral wall-to-wall covering, rather than the type of unfitted carpet which is sometimes known as an 'area rug'.)

As the pace of industrialization accelerated throughout the 19th century, technical innovations on both sides of the Atlantic allowed carpets to be produced in more complex designs and in greater volumes to reach a wider market. A similar watershed occurred after the Second World War with the advent of synthetic fibres and tufted carpet construction.

In the modern home, fitted carpet provides a fail-safe way to increase the sense of space where proportions are cramped or limited. Running the same carpet throughout a number of different rooms and adjoining areas has a seamless, coordinated effect, particularly if everything is on the same level. Alternatively, where space is not an issue, carpet can be very effective as a way of emphasizing a change of atmosphere. Carpeting in a bedroom, study or living room underscores the difference between busy, public or working areas and more relaxing private enclaves. In this respect it can create a sense of enclosure rather than spaciousness.

Choosing carpet demands careful research. Carpets vary in just about every conceivable way, from composition, construction and grade to colour, pattern and texture. All of these parameters will have an impact not only on appearance and performance, but also on your budget. In general, it is always better to opt for the best quality you can afford and this is even more true of carpet than of other types of flooring. Cheap carpet wears incredibly badly and will need replacing within a short space of time, which may well entail extra expense in the long run. If you are looking for a bargain, it is better to search out suppliers specializing in room-sized remnants, ends of rolls or discontinued stock than to make a compromise on basic quality.

In terms of style and appearance, the range of colours and designs offers virtually unlimited decorative potential. It is important to remember, however, that with any luck – and careful maintenance – a good carpet should last for many years, and may well have to work with successive wall colours or decorative schemes. On the other hand, plain neutral carpet may be safe but is unlikely to set pulses racing. The right carpet should offer some stylistic contribution of its own without setting unrealistic limits on what you can do with the room in the future.

Carpet styles tend to go in and out of fashion with perhaps rather more frequency than other types of flooring. One case in point is shag pile. The epitome of chic several decades ago, shag pile was subsequently cast aside as irredeemably downmarket and suburban, only to be rediscovered, more recently, by knowing young converts to 1960s style. Charcoal grey cord carpet enjoyed a long run of popularity among architects and designers who favoured its no-nonsense neutrality, but sheer familiarity has inevitably dulled some of its hard-edged impact. Today, stripy or muted flatweaves and tightly woven pale wool with cut-and-loop self-coloured relief patterns are experienced something of a vogue, in part in response to the contemporary interest in a natural look.

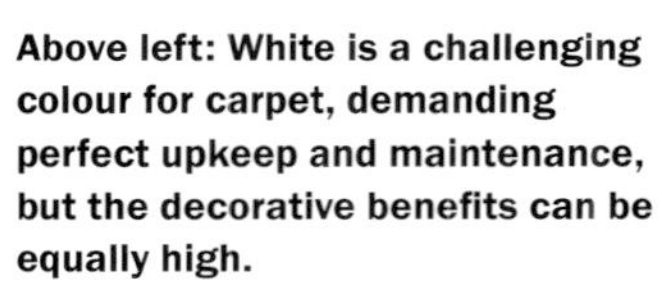

Above left: White is a challenging colour for carpet, demanding perfect upkeep and maintenance, but the decorative benefits can be equally high.

Top: A dark wood strip makes a neat edge between surfaces.

Above: Pleasing neutrality acts as a foil for strong architecture.

Right: Wall-to-wall carpeting in muted colours goes hand in hand with the contemporary look.

Practically speaking, there are various grades of carpet to suit most locations in the home, other than areas likely to get wet regularly. Carpet can be used in bathrooms; indeed, in many homes it has a luxurious and humanizing effect in what might otherwise be a cold and clinical room. But it is inadvisable to carpet a family bathroom, where spills and splashes are more common: carpet does not withstand repeated soaking and will eventually rot. Similarly, carpet does not represent a practical choice for kitchens, except for carpet tiles whose aesthetic appeal leaves much to be desired.

Carpet, as a soft floor covering, is resilient and quiet as well as warm and comfortable. It muffles sound and cushions feet. Deep- or shag-pile carpets can, however, be surprisingly tiring because of their extreme softness, in the same way as walking on dry shifting sand can be an effort.

Manufacturers have worked hard to counteract carpet's basic practical deficiency, which is its readiness to stain. Many carpets have a protective finish or can be treated to make them easier to maintain on a daily basis. However, there is no avoiding the fact that carpet does stain and demands a certain degree of vigilance to remain in good condition. Familiarity with stain-removal techniques is useful and it is a good idea to keep supplies of solvents and cleaners to hand for emergencies. Some people choose to solve the problem by opting for a hectic or busy pattern in the hope that it will not show dirt or stains; such designs do provide basic camouflage, but the overall effect of the floor may be unfortunately reminiscent of the hotel lobby or airport lounge.

Another disadvantage, for some households, is that carpet tends to harbour dust mites which can trigger allergic reactions and asthma. Cat and dog fleas are also rather fond of it and a serious infestation can be difficult to eradicate. Synthetic fibres, such as polypropylene, have been identified by the environmental lobby as posing a degree of health risk, while nylon can cause a build-up of static electricity.

Types and characteristics

Before choosing a carpet, familiarize yourself with the range of options available. You need to consider the fibre content, texture, construction, pile weight and density – not to mention colour and pattern. Don't be tempted to base your decision on appearance alone. Good underlay is also essential and will add years to the life of the carpet.

Grade and pile weight

Different grades of carpet are suitable for different locations. Some can take only light wear, as in bedrooms, while others will stand up to the heavy traffic of halls and stairs.

The resilience of a carpet is a function of its density, in other words how closely the fibres are packed together, and has little to do with the actual thickness or depth of the pile. To assess density, press the pile down with your thumb to see how quickly it springs back into shape; if it recovers almost immediately, it is dense and will be hard-wearing. You can also gauge the density of woven carpets by examining the back to see if the tufts are tightly spaced or by digging your fingers down into the pile to the base, then spreading the tufts apart to see if there is much room between them.

Below and right: Long or shag pile carpet is suitable only for light use and can be a little difficult to keep clean.

Far right: Short or velvet pile has a more seamless look and is luxurious.

There is no uniform standard of grading in carpet manufacture, but pile weights are usually given on the label and can provide a useful benchmark:

Light domestic use for example bedrooms, though not necessarily children's bedrooms, and other areas which are not heavily used: up to 800 grams per square metre.

Light to medium domestic use for example living rooms: 875 grams per square metre.

Medium to heavy domestic use such as family living rooms: 950 grams per square metre.

Heavy domestic use for instance stairs, landings and in halls – the connecting areas of the home – especially where there is a direct link to outside: 1 kilogram per square metre.

Carpet construction

The main types of carpet construction are woven, tufted and non-woven, although the method of construction is not necessarily an indication of quality. Non-woven types are generally cheap and often look it, but there may be little qualitative difference between well-constructed woven and tufted carpets of similar pile weight. There are also flatweave carpets, with no pile at all. Carpet is available in different widths – anything over 1.8m wide is termed 'broadloom'.

Woven carpets These have the pile woven along with the backing, which makes them strong, hard-wearing and generally fairly expensive. There are several styles:

Axminster carpet is made by inserting the pile into the backing from above then cutting it, a method which enables a large number of colours to be used since strands do not have to run along the back. (The term is confusing, since it does not refer to the products of the Axminster carpet factory, which was originally established in England in the middle of the 18th century to make hand-knotted pile carpets.) Axminster carpets are cut-pile and often highly (sometimes luridly) patterned. They are usually expensive and hard-wearing, and are often chosen for commercial locations.

Wilton carpet is named after the factory established in south-west England around 1740. The pile yarn is continuously woven into the weft, leaving loops, which can be left uncut, cut to make smooth cut pile or sculpted in a mix of cut and uncut loops for relief patterns. The method of manufacture means that fewer colours can be employed, normally only up to five (unlike the Axminster construction), since the different colours must be carried through the backing until required by the pattern. Wilton is more usually synonymous, however, with plain smooth-cut pile carpet of high quality and corresponding expense.

Flatweave carpets, as the name implies, have no pile at all. They are normally made of wool, or have a very high percentage of wool, and have become increasingly popular in recent years. Pale muted or natural colours are typical, as well as stripes or plaid patterns that run through the weave. The effect is tailored and discreetly elegant, with a similar aesthetic to the finer natural fibre coverings such as jute or sisal.

Tufted carpets These are the product of high-speed modern manufacturing techniques developed during the 1950s. The pile is inserted into the pre-woven base material by needles and may be left uncut in loops, sliced for cut pile or produced in a combination of the two. The backing is coated with adhesive to keep the pile in place and a second backing may be added for increased strength, sometimes incorporating a foam rubber underlay. Tufted carpets tend to be cheaper than woven varieties. Solid colours are standard, but flecked and simple printed patterns are also available.

Non-woven carpets At the cheapest end of the market are non-woven carpets, which are made by bonding pile fibres (usually synthetic) to the backing with adhesive, flocking the fibres electrostatically to the backing, or needle-punching them into the backing and sticking with adhesive. Some non-woven types are designed to resemble woven cord. In the domestic market, these carpets are generally thin and lacking in resilience; contract versions are very hard-wearing. They have no pile. Many carpet tiles are made this way.

Carpet pile and textural variation

Carpet texture is not solely affected by the fibres used – the pile makes an important contribution. Effects range from smooth plush velvety surfaces to self-coloured pattern in tight low loops. Some of the more common ones are:

Cut pile in which the pile loops are cut to make tufts of yarn that stand upright, is smooth and matt. Because of its smoothness, it may show footmarks or shading.

Velvet or velour pile is a very smooth and soft version of cut pile. It is quite hard-wearing and easy to clean, but again may show footmarks or shading.

Loop pile has uncut loops; the longer the loops, the bulkier and lighter the carpet. Patterns can be created by loops at different levels. Top of the range is Brussels weave, an expensive, hard-wearing form of loop pile in which the warp and weft yarns are equal in number. Cord is a tight low loop pile which resembles corduroy and is very hard-wearing. Berber is a loop pile made of natural undyed wool, although the term also applies to variegated or flecked wool. It makes a good all-purpose carpet.

Cut-and-loop pile, as its name implies, mixes cut-and-loop pile types to create relief patterns. It is normally available in solid colours only. Patterns, including herringbone, diamonds and basketweave, are made by cutting some loops and leaving others uncut at the same level, or at higher or lower levels.

Hard twist (or **frisé cut pile**) is a type of cut pile in which the fibres are twisted and set to give a tighter texture that does not fluff – nor does it show footmarks or shading. It is hard-wearing and ideal for stairs and areas of heavy traffic.

Shag pile refers to carpets with a pile up to 50mm long. Suitable only for light use, it should not be used on the stairs, where it may be hazardous. Shag pile is hard to keep clean and mats and tangles easily. The long pile can provide an attractive home for cat and dog fleas.

Right and far right: Customized carpet, made to the individual's own specification, allows exciting and original effects to be created. In this sleek modern interior a pathway of subtly contrasting colour has been created, which deliberately leads the eye through the space.

Below: Texture is an important dimension to consider when choosing a carpet. This shag-pile carpet in off-white adds an extra dimension of comfort to an intimate bedroom.

Carpet composition

Many different fibres, both synthetic and natural, used singly as well as blended, are currently used to make carpets. The fibre content is one of the most important factors determining the price, appearance and performance of a carpet.

The finest carpets have always been, and still are, all-wool. Yet the difference between natural and synthetic materials is not as great as it once was and the majority of carpets sold today are blends, with the synthetic component lending strength as well as bringing down the price to more reasonable levels. The most common fibres used in carpet manufacture are:

Wool is the classic carpet fibre. Nothing feels or looks as good as wool. Warm, soft and springy underfoot, it does not dent or flatten easily and is fairly simple to clean, as well as being anti-static and flame-resistant. The best choice for the environmentally conscious, wool is a natural material and an expensive one. Some of the finest carpet wool comes from New Zealand. A pure wool carpet is not particularly durable, however. An 80% wool, 20% nylon or 40% wool and 60% acrylic makes a more hard-wearing all-purpose blend. Pure or blended wool carpets will require moth protection treatment.

Linen shares the natural credentials of wool but has more limited applications in carpet terms. Linen is extremely expensive and far from hard-wearing as a floor covering, but it is elegant and unconventional. To maximize its performance, linen is best laid over thick underlay.

Acrylic is a synthetic fibre which bears a close resemblance to wool. It is, however, much more likely to be flattened by heavy furniture and to show the dirt; it also has poor flame resistance. Acrylic carpet is cheaper than wool, but generally more expensive than nylon.

Nylon is a very strong and durable synthetic fibre which is often used in blend with wool. In spite of its generally poor image, good-quality nylon carpet is very soft and available in a wide range of colours – and it can be expensive. Some nylon is treated to make it less static and stain-resistant types are also available. Cheap nylon, though, has a harsh texture and soils easily. All nylon melts rather than burns.

Polyester is a cheap synthetic which is soft, hard-wearing and easy to clean. It is not as springy as acrylic and is often used to make shag pile carpet. It has poor flame resistance.

Polypropylene is another cheap synthetic fibre which is being increasingly used in good-quality blends because it is very hard-wearing and stain-resistant. It flattens easily, though, so is not suitable for longer pile carpets. It is also flammable and does not take dye well.

Viscose is another cheap synthetic, which consequently tends to be used in poor-quality carpeting and which looks just that. It is also flammable and easily soiled.

Custom carpeting

Carpet is available in a vast range of colours, textures and patterns. Nevertheless, there is still a market for custom design — bespoke services exist to create unique carpets to individual specifications. If you want to commission a new carpet from a historical fragment or document or to devise an original pattern incorporating a specific motif, custom design is the answer. There are textile designers specializing in custom carpet who can also interpret a colour scheme or style for you. Such services are expensive.

Carpet tiles

At the opposite end of the market from custom designs, carpet tiles combine the appearance of carpeting with a practical format. They have a much wider application in commercial interiors, than in the home, and their aesthetic tends to reflect this. Carpet tiles come in a range of different-sized squares, fibres, construction and backings and in plain and patterned designs. They can be laid to resemble an unbroken expanse of carpet, or in various tile patterns combining different colours or designs.

The advantage of carpet tiles is that they can be easily and quickly laid and just as easily moved or replaced, making it possible to deal with patches of wear individually rather than go to the expense and trouble of re-covering an entire floor. They also allow easy access to underfloor services. In the home, they are perhaps most useful in areas likely to receive a great deal of wear and tear but where a soft surface is still desirable, such as in children's rooms.

Carpet tiles vary in quality and hence appearance: the better contract versions can be quite handsome, but cheaper non-woven mass market tiles are often fairly dreary. They can be stuck down or loose laid. Moving the tiles around from time to time will distribute wear evenly and is often a better idea than replacing a few worn tiles with conspicuously new ones.

Underlay

A good quality underlay will prolong the life of your carpet and make the floor warmer, quieter and more comfortable to walk on. Underlay compensates for any slight unevenness in the sub-floor and provides a barrier against dirt and moisture. Quality is determined more by the material than its thickness; very thick

Top: This cream carpet features a grid pattern formed from alternating loop and pile textures.

Above: Flatweave carpets have become increasingly popular. This example has a raised design.

Left: The white cord carpeting, matched with deep white skirtings and bed base, creates a great sense of light and expansiveness.

Opposite, far left: Wool is still a popular choice of carpet fibre; wool blends wear very well.

underlay is best avoided, since it can make the floor too soft and giving. All fabric-backed carpet requires underlay; foam-backed carpet should be laid on felt paper to prevent the foam backing from sticking to the floor.

Felt underlays are best used with quality woven carpets. These natural underlays consist of jute, wool or other animal hair, or a mixture of both. Foam or rubber underlays tend to be used under tufted rather than woven carpet. The best have a firm texture which does not crumble. These underlays should not be used on stairs, over underfloor heating or in damp conditions. There are also underlays which are a mixture of felt and rubber.

Choosing and buying carpet

Visit a carpet supplier or the flooring department of a good department store to compare different types and ask for samples to take home and live with for a while before committing yourself to a decision. Large retailers with big turnovers offer competitive prices; suppliers of brand-name carpeting tend to be more expensive. For bargains, try alternative sources specializing in room-sized remnants or ends of rolls – the range will be more limited, but the savings may be considerable. Many of these outlets sell carpet originally designed for heavy-duty commercial use, which means that larger remnants are often available.

To make accurate price comparisons you need to take accurate measurements of the area you intend to carpet (see page 167). Many suppliers provide a measuring service; all should be able to advise how best to minimize wastage and reduce the need for seams. Remember to include the cost of underlay, grippers and other accessories, as well as the cost of fitting.

Laying

Carpet should be laid on a dry, even sub-floor which has been properly prepared. Skirting boards can stay in place, but doors may need to be removed and planed to accommodate the increased height of the final floor. A well-laid carpet will last much longer than one which has been inexpertly installed. It is possible to install foam-backed carpet yourself (see pages 178-79), provided you are strong enough to manipulate the rolls, but laying fabric-backed carpet demands skill and proper tools.

Maintenance

Everyone knows that keeping carpet in good condition means frequent vacuuming. Dirt is not merely unsightly and unhygienic, it also works its way down into the pile, where it wears away at the fibres. Brushing or sweeping may tackle surface debris but only a vacuum cleaner can reach the base of the pile. Avoid unnecessary indentation by placing furniture cups under the legs of sofas and heavy tables. See pages 180–81 for stain-prevention of carpets.

Natural fibre coverings

Flooring made from natural fibres, including sisal, coir, seagrass and rush, has won many hearts in recent years, making the transition from the eccentric to the mainstream with remarkable speed. And yet, although such materials convey a fresh, contemporary image, they have one of the most ancient pedigrees of all types of flooring – a history of use that dates back centuries.

Long before our medieval ancestors covered the draughty stone floors of their castles and hall houses with a layer of disposable rush and straw, the ancient Egyptians were weaving mats from bulrushes. Until a couple of decades ago, however, most people were acquainted with these materials only as the ubiquitous coir doormat or the serviceable utility covering on the floors of old country houses. All that was before modern production techniques transformed natural fibre flooring into a serious alternative to carpet, available in wide widths that could be fitted wall to wall.

The renewed popularity of these products owes much to the modern desire to live with natural materials. Unlike the majority of carpet on the mass market, which contains at least some percentage of synthetic ingredients, natural fibre coverings are wholly natural, from sustainable resources, and affordable. They look authentic, too, with their natural neutral tones and nubbly woven textures, offering a pared-down simplicity that is also in tune with modern taste. Not that colours are limited to the muted biscuit shades of the basic ingredients; strong, rich shades and graphic patterns have been introduced to widen the decorative potential. Aesthetically speaking, these coverings are very versatile, easily

Right: Seagrass is the smoothest of natural fibres. It is also naturally stain-resistant – making it almost impossible to dye. The biscuit tones of the basic ingredient are, however, inherently pleasing and the nubbly weave gives it a dimensional character. Here, seagrass matting loose-laid over stripped floorboards, heightens the elegance of period decor.

Above, far right: Natural fibre coverings have found their way into every area of the home. Their understated neutrality provides a foil for contemporary or period decorative styles. Choose the right fibre for the location; some are more hard-wearing than others. Jute is particularly soft and is best used in areas of light traffic, such as bedrooms.

Far right: Coarser and more overtly textural, sisal is one of the most popular of the natural fibres. Here sisal matting serves as a base for a contemporary seating area.

Below, left to right, top to bottom: Jute (bleached); jute (unbleached); seagrass; coir (bouclé); coir (herringbone); seagrass (herring-bone); sisal (dyed bouclé); sisal (bouclé); and sisal.

ELLSWORTH KELLY

accommodating every decorative look – town or country, period or contemporary. Ecologically sound, characterful and adaptable, natural fibre coverings deserve their status as classics in the making.

Most natural fibres are reasonably hard-wearing, although jute is the least durable and will not stand up to heavy traffic. It is also considerably softer than the other natural fibres, for which reason it is the only one which approaches the comfort of carpet. The prickliness of coir and (to a lesser extent) sisal can be rather unpopular: children in particular find them rather inhospitable.

Natural fibre coverings are not as resilient underfoot as carpet, although using underlay improves the resilience. With the exception of rush, they are impractical for damp or humid areas such as bathrooms or conservatories, and also hard-working kitchens and utility rooms. Seagrass can be slippery and really coarse weaves of coir and rush are hazardous on stairs because heels can easily catch. By and large, most natural fibres shrug off dirt and debris, but they do stain. Protective treatment is advisable. All of these types of flooring are anti-static. Most are backed with latex, which prevents the accumulation of grit or dirt underneath, and makes installation simple; all can be laid wall to wall. Most types compare favourably in price with carpeting and, like carpeting, require professional fitting.

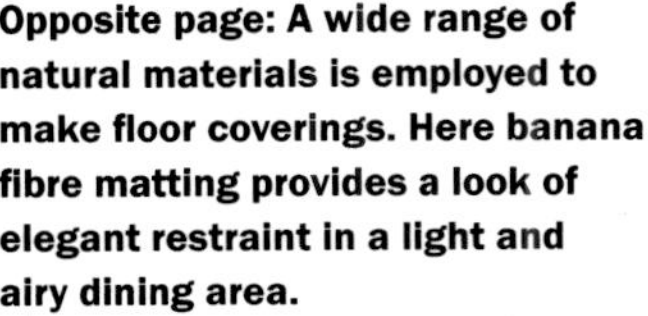

Opposite page: A wide range of natural materials is employed to make floor coverings. Here banana fibre matting provides a look of elegant restraint in a light and airy dining area.

Top: While there is a strong resemblance among the different natural fibre materials, there are some variations in practical performance. Seagrass is very durable and stain-resistant.

Above left: Slate-grey sisal is a handsome stair covering. Natural fibres should be professionally laid over special underlay.

Above right: A blend of neutrals: a fibre mat on a limestone floor.

Left: Natural fibre coverings can be used as loose area rugs as well as being laid wall-to-wall. As with any other rug, underlay is always a good idea and will improve wear.

Types and characteristics

Each type of natural fibre has distinctive attributes, in both practical and aesthetic terms. With the exception of rush, these floorings come with latex backing but the materials are also available in the form of mats, runners and rugs.

Seagrass is the smoothest of the natural fibres, which makes it more comfortable underfoot. The name refers to the fact that the crop is grown like rice in paddy-fields which are flooded with seawater during the growing season. Seagrass is tough, cheap and anti-static. Because the hard fibre is virtually impermeable, it does not stain like other natural fibres and shrugs off dirt. This characteristic, however, also means that seagrass cannot be dyed. Nevertheless, the basic 'natural' colour can be enhanced by incorporating coloured weft strings in black, green, red or blue.

Seagrass can be used almost anywhere in the home and can make a successful floor in kitchens, bathrooms and conservatories provided it is not subject to excessive wetting or spills. It should only be laid on stairs with the grain parallel to the tread, otherwise its smooth surface would be too slippery.

Coir is a fibre that derives from the coconut husk, from which it is beaten after the husk is soaked in fresh water. It is traditionally used for making doormats and sacking and, being naturally coarse and prickly, it is not especially comfortable for bedrooms or children's rooms. Very coarse, loose coir weaves are not suitable for stairs; heels will catch. Standard coir has an innately rustic look, but there are more sophisticated designs, with striped or chevron patterns in rich reds, blues and greens. Coir comes in a variety of weaves, including bouclé, basketweave, diamond and herringbone and, blended with sisal, it now comes in tile format. The standard version is very cheap and hard-wearing but it does stain.

Above: Sisal is easy to dye, which increases decorative possibilities.

Left: The nubbliness of natural fibres is all part of the appeal.

Far left: Seagrass enhances the Oriental mood conveyed by this burnished decor.

Right: Sisal matting can look surprisingly sophisticated.

Sisal is an exceptionally versatile fibre, in terms of its decorative potential and performance, and is accordingly the most popular of all the natural fibres. The basic fibre comes from the leaves of *Agave sisalana*, a dark-green spiky bush which grows in the subtropics and has a traditional use in making rope and twine.

The texture of sisal is mid-way between the hairiness of coir and the softness of jute, which makes it sufficiently hard-wearing to take heavy traffic but still acceptable for bare feet. It is easy to dye and comes in a variety of shades, patterns and weaves as well as in blends with wool. Inevitably, its readiness to accept dye means a corresponding lack of stain-resistance. Stain-inhibition treatments are advisable. Sisal is more expensive than coir or seagrass.

Right: Sisal, laid wall-to-wall as an integral covering, sets the tone for decor based around soothing natural themes.

Above, far right: Here the blond-on-blond effect is achieved by combining coarse sisal matting inset within an expanse of gleaming hardwood parquet — emphasizing the light, airy space.

Below, far right: Sisal matting, edge-bound, is loose-laid over sisal flooring for an interesting layering of texture.

Above: A loose-laid natural fibre mat defines a dining area. The shift underfoot from wooden floor to matting is subtle, owing to the fact that the materials are tonally very similar. The textural variety provides a sense of character without detracting from spatial quality.

Left: Sisal is easy to dye, which means that it is available in a wide range of soft natural colours and patterns. Designs include stripy runners and bold chevrons, as well as more pictorial effects, such as this 'needlepoint' sisal matting.

Jute comes from a plant native to subtropical regions of India and has been exported to the West since the 18th century for use in rope-making and as carpet backing. The fibre is stripped by hand from the stalks of the plant which have been softened in water. Jute is slightly cheaper than the best sisal, but far less durable. It is soft enough to be more than welcome in the bedroom, but will not stand up to heavy wear and requires stain protection. It is available in a range of elegant weaves and in natural, and bleached tones, as well as pastel and rich colours.

Rush or medieval matting makes a heavy, robust flooring that can be used in a wide range of locations. It is inherently countrified in appearance, yet it can also make a stylish contemporary floor. The matting is made of hand-plaited strips of rush, each 9cm wide, which are sewn together to make a room-sized covering (which is laid loose) or mats of any dimension. Medieval matting is at the top end of the natural fibre price range. It should not be used on the stairs because it is too smooth and slippery. Rush matting requires regular sprinkling with water to maintain its condition and makes a good flooring for naturally damp areas such as conservatories and bathrooms provided there is adequate ventilation.

Laying

Like carpet, natural fibre coverings require a dry, level, even sub-floor which has been properly prepared. Underlay is not essential, but it will serve to correct any slight imperfections in the sub-floor and provide extra resilience and wear-resistance. However, if you intend to use these materials on stairs, underlay is always a good idea as it will make the covering much more hard-wearing.

Before installation, natural fibre coverings should be laid out in the area where they are to be fitted for 48 hours in order to acclimatize the fibres to the ambient humidity levels. Latex-backed natural fibre

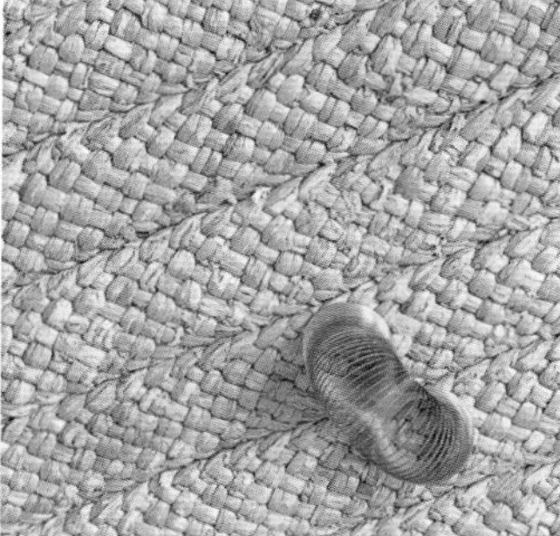

Top and above: Rush matting tends to be rather more expensive than other types of natural fibre coverings and needs regular sprinkling with water to maintain its condition.

Above, right: Absence of strong colour need not mean absence of vitality, provided textures are contrasted for depth of character.

Left: Rush matting retains something of a rustic aesthetic. Here it is laid in a chequerboard pattern in a country hallway.

Far left: Natural textures combine effortlessly. Here a felt rug is laid over natural fibre matting.

coverings should be stuck down all over, either directly to the sub-floor or to the underlay which has been stapled to the floor. Fitting is professional work. The recommended adhesive leaves no sticky residue, which means the covering can be taken up at a later date without damage to the sub-floor. Seams can be tightly butted up or taped. Gripper rods around the perimeter of the room in addition to the adhesive also help to prevent subsequent shrinkage.

Maintenance

Many natural fibre coverings are susceptible to wear from castors, so it is best to have additional protection under chair and sofa legs. High, spiky heels are also best avoided. Bright sunshine may cause fading in some varieties.

Natural floor coverings react differently and usually (with the exception of medieval matting) adversely to humidity. Coir expands with moisture and may wrinkle and buckle. As it dries out, joints may open up. Sisal, on the other hand, shrinks when wetted. In very damp conditions all natural fibres will rot.

For regular maintenance, all of these floorings should be vacuumed. Never wash or shampoo. Coir, sisal and jute are best treated with a stain-inhibition treatment. Tackle spills immediately, before staining has a chance to occur. Let muddy tracks dry out and then brush along the grain and vacuum.

Rugs

Superficially, rugs are finishing touches, the decorative flourishes that make rooms look better-dressed. However, both practically and aesthetically rugs can be a key element in making a room complete, rather than merely optional accessories.

Rugs domesticate floors. They provide an essential layer of comfort, warmth and sound protection that helps to mitigate some of the disadvantages of hard flooring, such as stone, hard tiles and wood. Rugs enable you to have your cake and eat it too, offering a soft surface to meet bare feet by the bedside or a cosy place to curl up in front of the fire without the need to carpet throughout the entire space. Unlike other types of flooring, rugs do not commit you to a permanent arrangement. They can be moved from home to home or room to room; brought out in winter and rolled away in summer.

Visually, rugs are no less defining elements. A rug transforms a sofa and chairs into a seating circle; a series of rugs in an open-plan space gives what could be just so much empty floor area a sense of enclosure and proportion. Rugs bring depth and character to decorative schemes, adding vitality to neutral backgrounds and providing a coordinating element where rooms are more richly coloured or patterned. A beautiful rug can serve as a room's signature, provide the foundation for the entire decorative approach, as well as pleasure in its own right.

Below: Rugs which display the soft colours of vegetable dyes can be layered for a sympathetic blend of patterning.

Rugs or carpets were once the treasured possessions of the privileged few. Up until the 18th century, fine carpets were never laid on the floor, but draped over tables or hung on walls. Today,

Above: Contemporary rugs make floor-level art. Bold geometric or abstract designs have great vitality.

there are many rugs that still merit this degree of reverence, glorious antique collector's items costing a king's ransom. But the majority of rugs now available, even Oriental or Persian ones, are within the reach of the average household, and many flatweave rugs are positively cheap.

For centuries, carpets were synonymous with rugs of Oriental or Middle Eastern provenance, normally fine knotted examples with a pile surface. Knotted carpets have been produced in various regions of the world since antiquity, but they only began to appear in the West in large quantities after the establishment of reliable trading routes during the 16th and 17th centuries. By the

Above: A new departure in soft flooring, rugs made of paper twine have a crisp, graphic look, ideal for contemporary settings.

18th century, the first factories producing a similar type of knotted carpet were established in Europe, and the spread of influence then became two-way – Middle Eastern and Oriental carpets designs began to reflect European taste. This cross-fertilization of ideas between East and West is no less marked today, with modern designers seeking inspiration and buyers from large retail chains in Europe and North America commissioning carpets, dhurries and kelims direct from local manufacturers in designs and colourways most likely to appeal to their Western customers. The results are often pleasing abstractions of ethnic patterns and motifs which lack the more overtly tribal appearance of some traditional designs.

The past few decades have seen an explosion of interest in more humble ethnic weaves, such as kelims and dhurries, flatweaves with naive geometric patterns. These eminently affordable rugs have a contemporary quality and freshness which works well in modern interiors. At the same time, the current crafts revival has brought the work of modern rug designers and makers to the fore, and a wide range of original rugs can be found, as one-off commissions, in limited runs or in mass-market quantities. All in all, there is more scope for floor-level art than ever before. Prices vary enormously. Simple cotton rag rugs cost no more than an average restaurant meal, while a good-quality Persian might set you back as much as a couture evening dress – and more. Rather than compromise on quality, it is often better to opt for a different type of rug altogether: what would not buy you a very exciting Oriental might well buy you a superb kelim, for example.

Types and characteristics

Rugs and carpets vary in construction and material as well as in provenance, colour, pattern and design. The terminology can be confusing: 'Oriental', for instance may be loosely applied to any carpet from the Near, Middle or Far East. At the upper end of the market, rugs are a subject for the connoisseur, with variations of design classified according to region or tribe as well as period. At the other end are the simple home-spun styles made from scraps of cotton or wool, originally out of necessity, of course, using whatever was to hand.

Knotted carpets, with a pile surface, include various kinds of Turkish, Persian and Chinese rugs. The quality of the carpet is a function of its density, with the finest having as many as 1,000 knots per square inch. Most of these carpets are made of wool, but very luxurious versions are made in silk.

Chinese carpets are thick knotted carpets made from wool or silk, often in light, clear colours, typically blue, yellow, peach and white. The motifs are fairly widely spaced and often feature stylized flowers, birds, butterflies, dragons or figures; and the surface of the pile is sometimes cut for a sculpted relief effect. This type of 'Oriental' rug is now made not only in China – Iran and India produce fine examples – and the best are expensive.

Tibetan carpets come from the remote regions of Tibet and Nepal. They are hand-knotted in wool, and can be of very high quality and of corresponding expense. Traditional designs and colours may show a classic Chinese influence, although rugs commissioned by Western buyers display a blend of classic and contemporary motifs.

Turkish carpets is loosely employed to refer to a range of different types of carpet, hand-or machine-made, from the Near and Middle East, with much of the production centring on the Anatolian region of Turkey. They are distinguished from Persian carpets by the type of knot used for the pile, known as the Ghiordes knot, which is tied symmetrically on two adjacent warp threads. Typically, designs are stylized nature motifs, sometimes with central medallions; there are also prayer rugs, featuring an arch whose apex is to be pointed in the direction of Mecca. Wool is sometimes mixed with coarser goat hair. During the latter part of the 19th century, the use of chemical or aniline dyes resulted in harsh colouring, but modern

Turkish rug producers are now returning to the use of traditional vegetable dyeing. Some weaving cooperatives are collaborating with Western designers and suppliers to produce modern designs with an ethnic flavour.

Persian carpets represent the height of weaving artistry. Technically, they are distinguished by the use of the Persian knot, in which yarn is twisted around one warp thread and under the adjacent thread, enabling the construction of a very dense pile; each rug represents hours and hours of intensive work. Made of wool or silk, patterns are rich and detailed, with stylized or naturalistic flowers, leaves, animals and birds scattered over the ground. Some designs recall a garden of paradise; others feature the tree of life, prayer arches and central medallions. The *gul*, a stylized flower motif in the form of a rounded octagon also known as an 'elephant's foot' and the *boteh*, a stylized leaf (similar to the motif in a paisley pattern), are also typical. The finest Persian carpets ever made date from between the 16th and 18th centuries and are now museum pieces; quality control has helped to keep modern standards high. True Persian carpets are made in Iran and other parts of central Asia, and are often classified by specific town or region of origin, such as Isfahan and Tabriz. The best are hand-knotted, but there are also machine-made versions from small factories, as well as reproductions made in Pakistan, Bulgaria and the West (sometimes unscrupulously passed off as originals). Background colours are often deep red or blue. Good-quality Persian carpets improve with age and should last for generations.

Turkoman or Bokhara carpets are wool rugs sourced from a wide area across central Asia, including Pakistan and Afghanistan and Turkmenistan. Many of these rugs feature the characteristic *gul* motif. Colours are typically dark, rusty red, black and blue. Bokhara, an important trading centre in central Asia, has become the general term for carpets woven by nomadic peoples of the region.

Caucasian carpets include a variety of different types of carpet, from knotted pile to flatweave, which derive from the Caucasus region between the Black Sea and the Caspian Sea. Motifs are stylized and often naive, with a graphic use of bright colour and intricate borders. Some designs are similar to very simplified Persian patterns; *soumak* are flatweave carpets with threads left loose at the back.

Far left, above: Good quality carpets are coloured using natural vegetable dyes, and hand-knotted.

Far left, below: Rugs are decorative centrepieces and provide a good way of defining space. This fine oriental rug adds elegance to a dining area.

Centre, left: In living areas, rugs placed centrally can work hand in hand with furniture placement to create relaxed conversation areas, drawing people together.

Left: Rugs on bare boards must be laid over non-slip matting.

Below: Rugs enrich interiors with pattern and colour and provide vivid focal points. Here, elements of the decorative scheme have been inspired by some of the colours displayed in the rug, to create a richly layered yet unified effect.

Kelim is the generic Persian term for a coarse woollen flatweave rug and is used today to refer to carpets hand-woven by nomadic peoples in the Middle East (specifically Afghanistan and Turkey) and North Africa. Colours vary, but rich reds, golds, creams, blacks and blues are typical; designs vary too, but are generally geometric and bold. Kelims are available in a wide range of sizes, including runners, and in different qualities. Good antique kelims are the most expensive, although they are still reasonably priced, compared to the finer Persian carpets. Kelims are not as durable as knotted pile carpets.

Dhurrie is the Indian version of the kelim, a flatweave rug usually made of cotton. Colours are often much less intense – almost sombre, compared to kelims – with the pale tones of grey, pink, ochre and light blue being typical. Although dhurries have been woven since antiquity, it is only in relatively recent times that these rugs have become widely popular in the West. There has been an exchange of colour ideas and designs between Western retailers and Indian producers: simple stripes, checks, zigzags and chevrons predominate. Dhurries do not provide much in the way of warmth or sound insulation, but their crisp, clean designs are much in tune with modern taste. They are extremely affordable, reversible and some can be machine washed, although shrinkage is likely.

Above: Flatweave carpets, such as kelims and dhurries, are at the more affordable end of the price spectrum. The pale tones of this traditional rug suit the simple modern setting.

Far left: The shock of the new: wool rugs in brilliant colours. The crimson of this tufted rug makes a stylish clash with orange upholstery.

Left: Vegetable dyeing produces rich, glorious colour without stridency. This traditional method of dyeing has one further advantage, which is a particular asset when it comes to patterned rugs. This is the fact that when the colours do fade, they fade in synchronicity with each other so that, although muted, tonally the design remains the same.

Right: Rugs are all about comfort. In the case of faux fur and sheepskin, that comfort has a distinctly luxurious, sensual dimension.

Aubusson carpets are tapestry-woven smooth-faced rugs, which take their name from the products of the original Aubusson factory established in France in the 18th century. Designs are typically feminine and floral, in light, pretty colours.

Needlepoint rugs, traditionally worked in tent stitch on a canvas backing, have the appearance of tapestry, with widely spaced floral motifs on a plain pale or dark ground. Handworked needlepoint rugs were common in early American homes, the product of hours of painstaking toil; machine-made reproductions are now available.

Serape is the name for a thin flatweave rug or blanket from Mexico and the southwestern United States, woven in searingly bright colours and often in striped patterns.

Shag-pile rugs include the woolly Greek *flokati*, normally white or off-white, and the Finnish *rya*, a long-pile rug in strong contemporary designs and bright colours.

Felt rugs include the *numdah* from Kashmir, which is a cheap, soft non-woven carpet often featuring naive motifs embroidered on a pale ground. Nursery versions with simple animals and alphabets are popular. Numdahs wear out quickly, show dirt very readily and don't clean easily.

Rag, hooked and braided rugs are examples of simple folk weaves which were originally made – out of necessity – from scraps of cotton material or old clothing, roughly stitched or looped together, hooked through a canvas backing or braided into plaits and sewn. Rag, hooked and braided rugs define a certain kind of countrified

style which is Scandinavian and North European in origin, but also strongly associated with the homespun look of colonial American interiors. These rugs are not difficult to make, albeit time-consuming, but mass-market versions are widely available and extremely cheap. Antique originals, on the other hand, are collectors' items.

Slightly more sophisticated are flatweaves in cotton or wool, often in soft natural colours and subtle textured patterns. Classic designs include stripy runners, a feature of traditional Scandinavian interiors, where they were often laid around the perimeter of the room to make a carpeted walkway over bare boards. Some of the colours used in modern interpretations are also inspired by their Scandinavian origins. Runners are also highly suitable for hallways and stairs, as they can be turned periodically to distribute wear evenly.

Right: Blues make a serene combination in complementary patterns. The stripy design of the flatweave rug, checked cushion covers and webbed seat ring the changes where colour is restricted.

Below: A complementary blend of stripes: stripy wool-pile stair carpet and flatweave wool runner.

Above, far right: A woven cotton rug laid over fine jute is the essence of pure style.

Below, far right: Flatweave cotton rugs make versatile — and washable — coverings for all areas of the home.

For modern rugs, crafts councils or other organizations promoting the work of new designers and craftspeople can be a useful starting point if you are looking for a one-off original. Otherwise, high-street retailers and department stores increasingly stock a wide range of rugs of all kinds, both traditional and contemporary, and the volume of sales generally guarantees sensible pricing. The range of stock generally includes 'area rugs' (essentially unfitted loose-laid carpet) and machine-made reproductions of traditional Persian, Oriental or Turkish carpets as well as the genuine articles. Some specialist suppliers produce reissues of classic patterns by famous designers such as William Morris or early modernists such as Eileen Gray. Natural fibre floorings are also available in the form of runners, rugs or room-sized mats from the same type of outlets that supply fitted natural flooring.

Choosing and buying rugs

In many areas of the world, the carpet-dealer has long been virtually synonymous with trickster or con-merchant. Buying Oriental rugs, antique or new, hand-knotted or machine-made, still demands a degree of expertise and wariness on the part of the purchaser. There is no real substitute for a working familiarity with different types and qualities of rug. If you intend to make a serious investment, it is best to study the subject in some detail and then take a long hard look at what is available from different sources before making a final choice. Real quality is normally more or less self-evident. A fine rug has a certain luminous look. The design will be crisp and clear, rather than blurred, the colours rich, not muddy. Rugs produced in areas of the world where the light is strong are

Below, left and right: Relief textures introduce a new dynamic to neutral shades. The effect is similar to other more traditional self-coloured patterns such as damask, but with a fresh, contemporary edge that lends wit and flair to a modern interior.

Left and above: A beautiful rug, whether contemporary or traditional, can be the signature of a room's decoration and style. Here the soft colours of the rug are picked up in the upholstery and furnishings in slightly stronger tones, while the loose abstract design creates a counterpoint to the strong modern lines. Pale hardwood flooring makes a discreet background surface.

Below: Contemporary art rugs can be commissioned directly from designers and craftspeople, or acquired through a craft gallery or similar outlet. A less expensive proposition are the modern rugs available from high street retailers which share a similar aesthetic. This careful composition in blocks and bands introduces discreet colour to basic neutral decor.

Far left, left and below: There is a wide variety of contemporary rugs on the market, from mass market to designer examples. For the ultimate custom rug, some companies will make up a design as a one-off. Always lay rugs over underlay to promote wear-resistance and decrease the risk of slipping.

Right: A circular rug with a bold radial pattern makes a great focal point. Such eye-catching designs are best laid over a neutral backdrop — they need a bit of breathing space to be fully appreciated.

Far right: Many contemporary rugs feature relief patterns that add textural variety to the basic pattern and colour mix.

often quite bright to Western eyes, but the colours will mature and soften with time. Rugs made of wool which has been vegetable-dyed age gracefully, the colours fading gently in synchronicity with each other, in contrast to chemically dyed products which retain a certain harshness.

Good sources to try include established and reputable dealers who have a long track record in the business, major retail outlets and, if you know what you are looking for, auctions. At all costs avoid so-called bankruptcy sales or special 'closing down' offers. Buying a rug in its country or region of origin is not a guarantee either of quality or savings: many of the best examples go straight to the export trade, leaving what are often little more than tourist souvenirs in the local markets.

Price is determined by basic quality (such as density, pattern and material), age and condition. If you do not mind small signs of wear, such as a bit of fraying or a ragged fringe, you may be able to pick up what is essentially a fine carpet at a bargain price. At the same time, it is important to beware the artificially 'aged' carpet, distressed by bleach or other means to simulate an antique, but which actually reduces their life. A good dealer will be only too pleased to provide evidence to support a rug's provenance.

Laying

All rugs placed on hard floors should be laid over a non-slip mat to prevent accidents. The extra padding provided by the underlay will also protect the underside of the rug from wear and add softness and resilience. Light rugs can be secured by mesh backing or bonding strips. An underlay pad is also useful if the rug is to be laid over carpet: it will prevent the rug from wrinkling and creeping and also protect the carpet from staining if the rug is not colourfast.

Practicalities

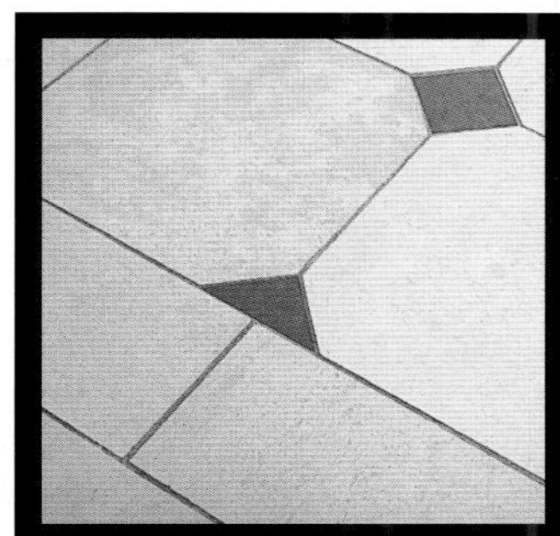
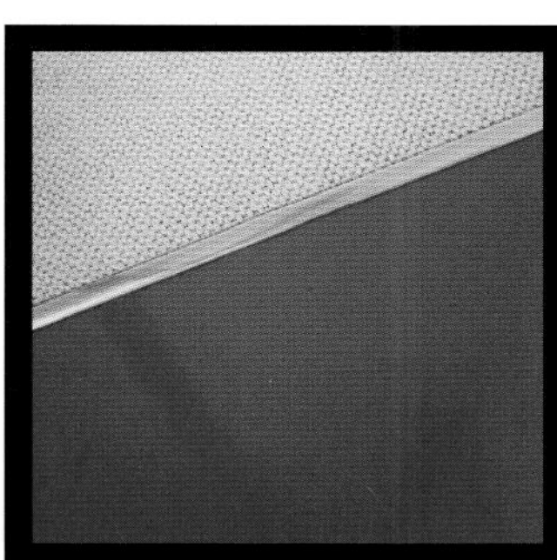

As the preceding chapters have indicated, laying a new floor tends to be a job for the professionals. Unlike other areas of home decoration and improvement, where the skills are relatively simple to acquire and the materials easy to manage, floor laying, with a few exceptions, is considerably more demanding. To begin with, more physical strength is usually required: wood, stone, sheet flooring, carpet rolls and the larger hard tiles are heavy and unwieldy: some back-breakingly so. Hard materials are often brittle, easy to chip, sheer, crack or otherwise damage if clumsily handled. They may well be expensive, too, which makes breakages all the more heart-rending. A range of specialist tools may be necessary to cut, fit and finish the floor, not to mention the specialist skills that only come with experience. Most floors present at least some obstacles and irregularities – such as pipework, alcoves or curves – which demand careful planning to circumvent. Floors composed of multiple units, such as tiles, particularly if these are to be laid in patterns, also need to be properly set out for a successful result. In most cases, mistakes cannot easily be rectified and a badly laid floor may wear out quickly, be hazardous or simply be an eyesore.

For all of these reasons, it is best to err on the side of caution and leave it all to the experts. However, if you are reasonably confident of your DIY skills, some types of floor laying are fairly straightforward and instructions for these are included in this section. Even if you do not intend to lay a new floor yourself, there are a number of practical issues that you will need to understand in order to choose the right flooring, buy the right quantity and commission the work properly.

Know your floor

At the outset, it is important to understand the distinction between floors and flooring. A floor is part of the structure of your home, whereas flooring is the covering applied to the floor. In some cases, floor and flooring may be one and the same thing: timber floorboards or old flagstones, for example. Most new floors, however, consist of some kind of flooring applied over a structural floor.

There are two main types of floor: solid and suspended. Solid floors occur at ground or basement level and are sometimes known as 'direct-to-earth' floors. Most suspended floors are made of timber and consist of two elements: the floor joists or beams, which span the walls, and floorboards or chipboard sheets laid across the joists. The type of floor structure will determine which floor coverings are suitable. Solid floors and suspended concrete floors can bear more weight than suspended timber floors. If your chosen flooring is very heavy, you may not be able to use it over a suspended timber floor at all. If you are in any doubt about the load-bearing capacity of the floor, it is advisable to call in a surveyor who will be able to do the calculations for you.

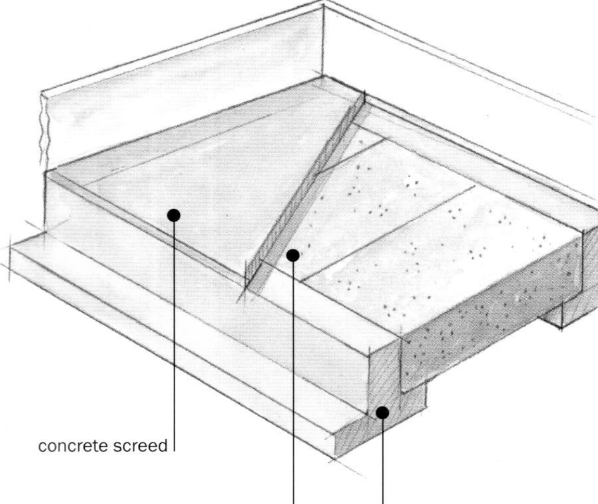

Solid suspended floors
In modern high-rise blocks and some new homes, suspended floors may be made of breeze blocks supported by concrete joists and covered with a concrete screed.

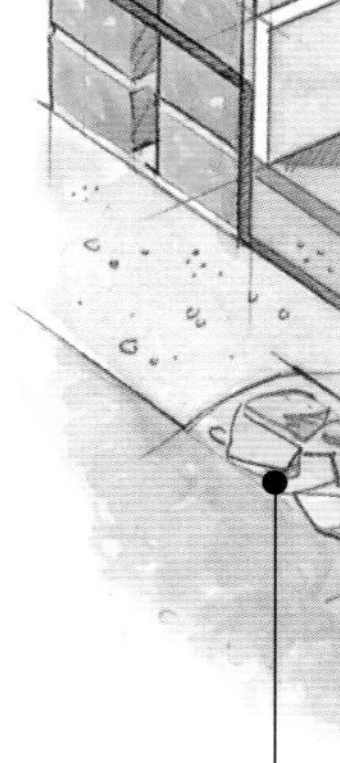

Solid floors
In homes built in the last fifty years, ground floors usually comprise a concrete slab, laid over a hardcore base and covered with a sand and cement screed. There should be a damp-proof membrane between the hardcore base and the concrete slab to prevent rising damp. (Older houses may have flagstones, brick or hard tiles bedded into the ground; such floors may not be damp-proofed.)

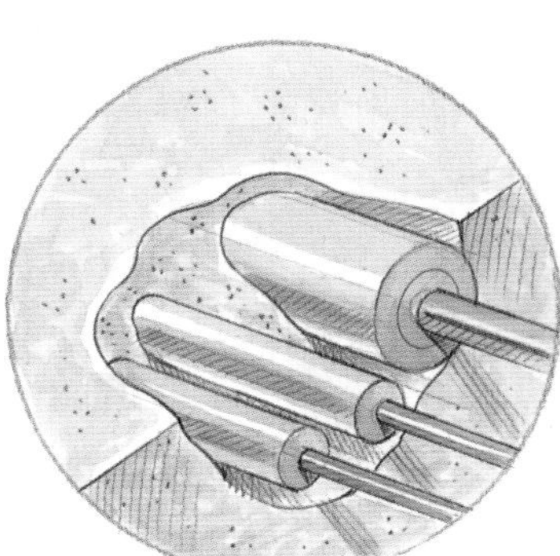

Underfloor services
Pipes and wires can be unseen hazards when new floor coverings are fitted. The danger is usually greater in the case of suspended floors; in solid floors, servicing and cabling are usually sunk some distance from the surface.

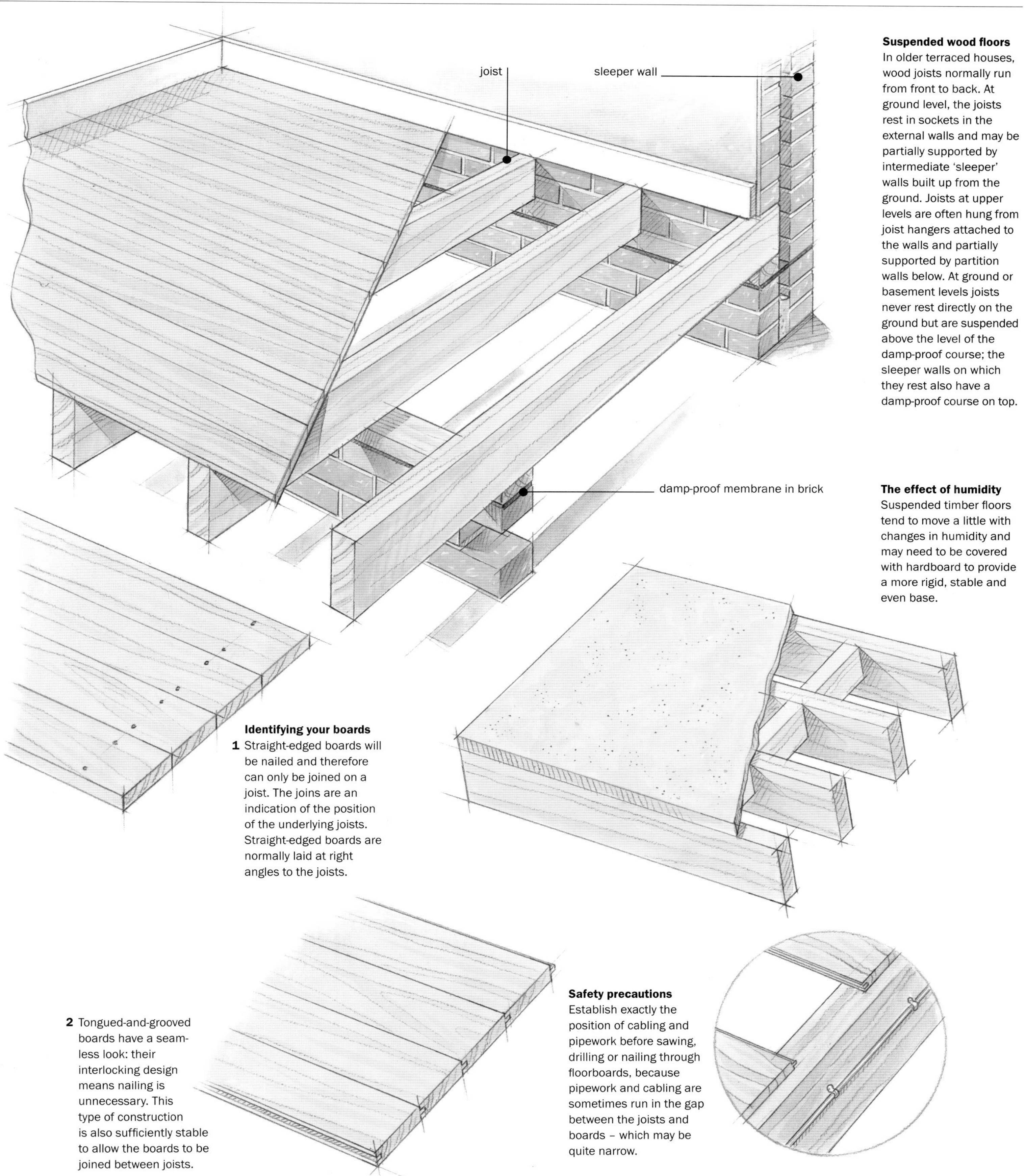

Suspended wood floors
In older terraced houses, wood joists normally run from front to back. At ground level, the joists rest in sockets in the external walls and may be partially supported by intermediate 'sleeper' walls built up from the ground. Joists at upper levels are often hung from joist hangers attached to the walls and partially supported by partition walls below. At ground or basement levels joists never rest directly on the ground but are suspended above the level of the damp-proof course; the sleeper walls on which they rest also have a damp-proof course on top.

The effect of humidity
Suspended timber floors tend to move a little with changes in humidity and may need to be covered with hardboard to provide a more rigid, stable and even base.

Identifying your boards

1 Straight-edged boards will be nailed and therefore can only be joined on a joist. The joins are an indication of the position of the underlying joists. Straight-edged boards are normally laid at right angles to the joists.

2 Tongued-and-grooved boards have a seamless look: their interlocking design means nailing is unnecessary. This type of construction is also sufficiently stable to allow the boards to be joined between joists.

Safety precautions
Establish exactly the position of cabling and pipework before sawing, drilling or nailing through floorboards, because pipework and cabling are sometimes run in the gap between the joists and boards – which may be quite narrow.

Remedial work

The condition of the sub-floor is of critical importance. Before new floor coverings are laid, the sub-floor must be inspected for any signs of damage or decay. Failure to treat deficiencies will result at best in a final floor which looks shoddy and wears badly; at worst, the structural fabric of your home may be weakened. Defects in solid floors, if they are basically sound, dry and level, can usually be made good by an amateur. Any damp must be cured or the floor covering itself will either not adhere or begin to rot. The source of the damp must be corrected, or the problem will only recur.

Timber sub-floors may also suffer from damp or infestation. If the floor feels bouncy or spongy the joists may be damaged. Wet rot shows up as soft dark patches, easily penetrated by a sharp tool, or cracks and splinters. Dry rot may show as musty grey threads or fungal growths. A peppering of fine holes indicates woodworm. In each case professional treatment is imperative. The affected timber must be cut out and the remaining areas sprayed with the appropriate chemical.

Remedying an uneven solid floor
A scratched or pitted surface can be filled by pouring over levelling compound. This is also the best way to smooth small ridges. Work with small quantities at a time and smooth the surface with a steel float.

Filling holes

1 Isolated holes or dips can be filled with cement mortar. Apply a concrete sealant to the sides of the hole first, to give a good key. Concrete sealant can also be applied over an entire concrete floor to prevent it 'dusting'.

2 Use a flooring trowel to get the repair as smooth as possible. Once dry, any remaining bumps or slight ridges can be rubbed down with a carborundum block.

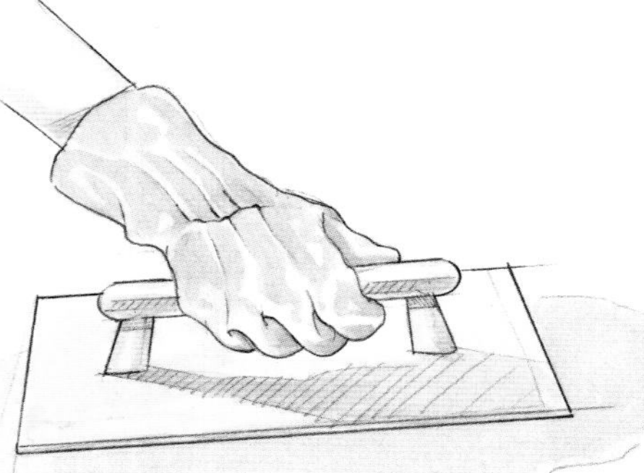

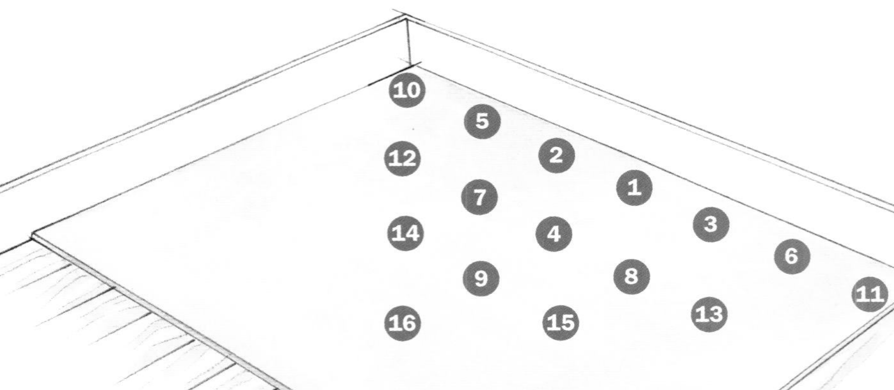

Correcting uneven boards

1 Use hardboard to cover the floorboards. Lay the sheets flat for 48 hours in the room to acclimatize to the humidity levels. Nail them, rough side up (to give more key for adhesives), at 6-inch (15 cm) intervals, using special hardboard nails, in a pyramid sequence to prevent the sheet bulging.

2 For a more solid base, use 6mm plywood rather than hardboard. For soft flooring (carpet, vinyl, rubber, etc), allow for access to underlying pipes and wiring before laying large sheets and ensure that no nailheads protrude above the surface, which would mar the look of the final floor. Heavy materials, such as mosaic or tile, may need 12mm exterior grade or treated marine plywood which will not absorb water from the adhesive or grout.

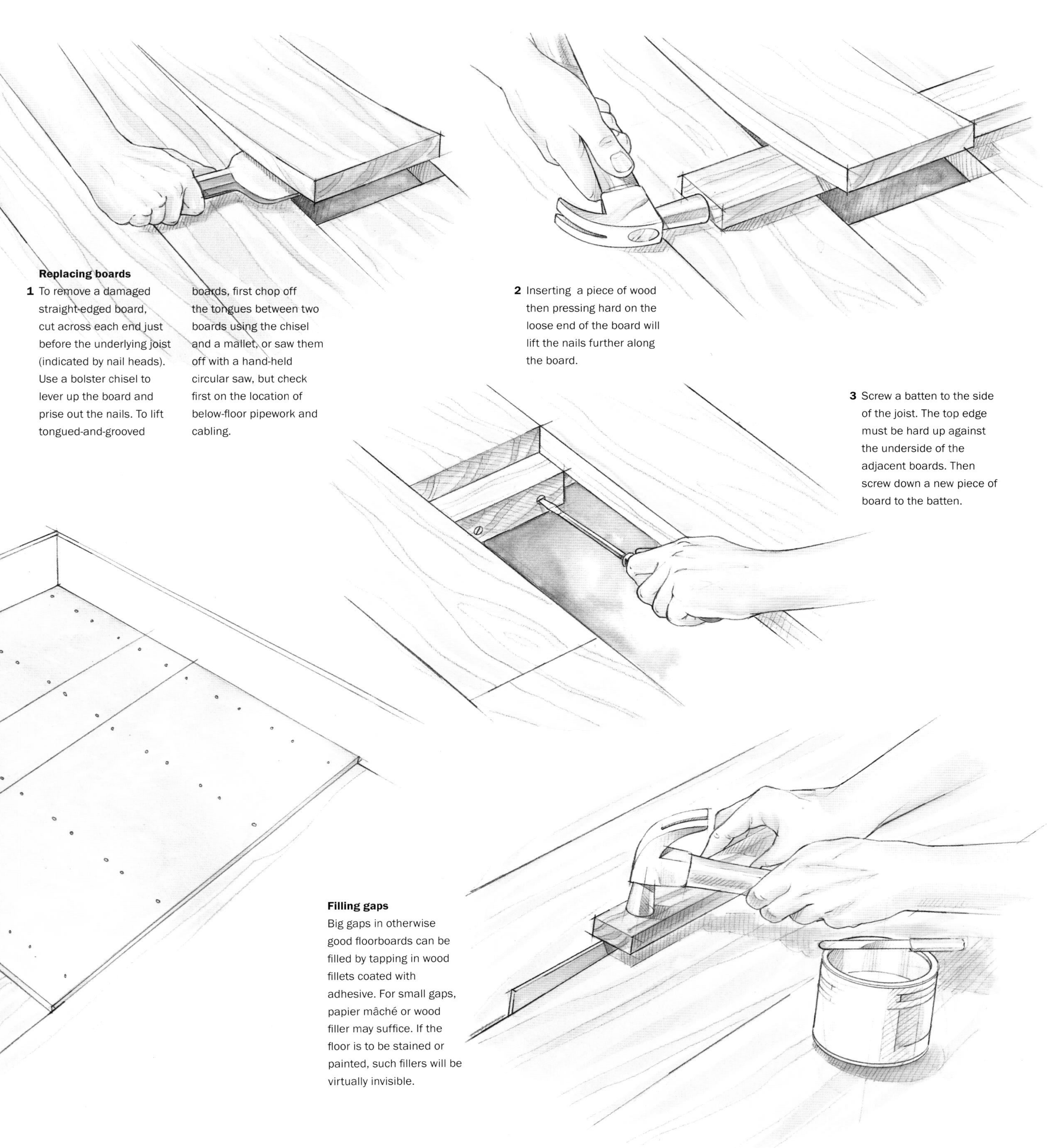

Replacing boards

1 To remove a damaged straight-edged board, cut across each end just before the underlying joist (indicated by nail heads). Use a bolster chisel to lever up the board and prise out the nails. To lift tongued-and-grooved boards, first chop off the tongues between two boards using the chisel and a mallet, or saw them off with a hand-held circular saw, but check first on the location of below-floor pipework and cabling.

2 Inserting a piece of wood then pressing hard on the loose end of the board will lift the nails further along the board.

3 Screw a batten to the side of the joist. The top edge must be hard up against the underside of the adjacent boards. Then screw down a new piece of board to the batten.

Filling gaps

Big gaps in otherwise good floorboards can be filled by tapping in wood fillets coated with adhesive. For small gaps, papier mâché or wood filler may suffice. If the floor is to be stained or painted, such fillers will be virtually invisible.

Measuring up

Accurate measurement is essential when ordering new floor coverings. Most people worry about underestimating the amount of flooring required; in fact, overordering is far more common and results in needless waste and expense.

Start by making a scale plan of the area in question and draw it up on graph paper, marking in the position of doorways, windows, fireplaces, alcoves, curves or any other irregularities of contour: rooms are rarely perfectly true in shape. A scale of 1:20 will provide enough detail. Choose whether to work in imperial or metric measures and stick to the same system for all your calculations: don't measure in one and then convert to the other when ordering – inaccuracies are bound to creep in. It is best to work in the system in which the material you wish to buy is commonly sold.

If you do not feel confident about taking the measurements yourself, many retailers, suppliers and fitters offer a free service and can advise on the quantities you will need, position of seams, and so on.

Estimating unit quantities For types of flooring sold in units, find out in what dimensions the material is sold. If you intend to use 10-inch square tiles, for example, divide the total floor area by the area of one tile to calculate the total number of tiles (plus a 5% wastage allowance). The same applies to boards, woodstrip, stone slabs, bricks, etc. Some mass-market flooring is sold in packs, with each pack covering a specified area. In this case, simply divide the total floor area by the area covered by each pack to determine how many packs you need.

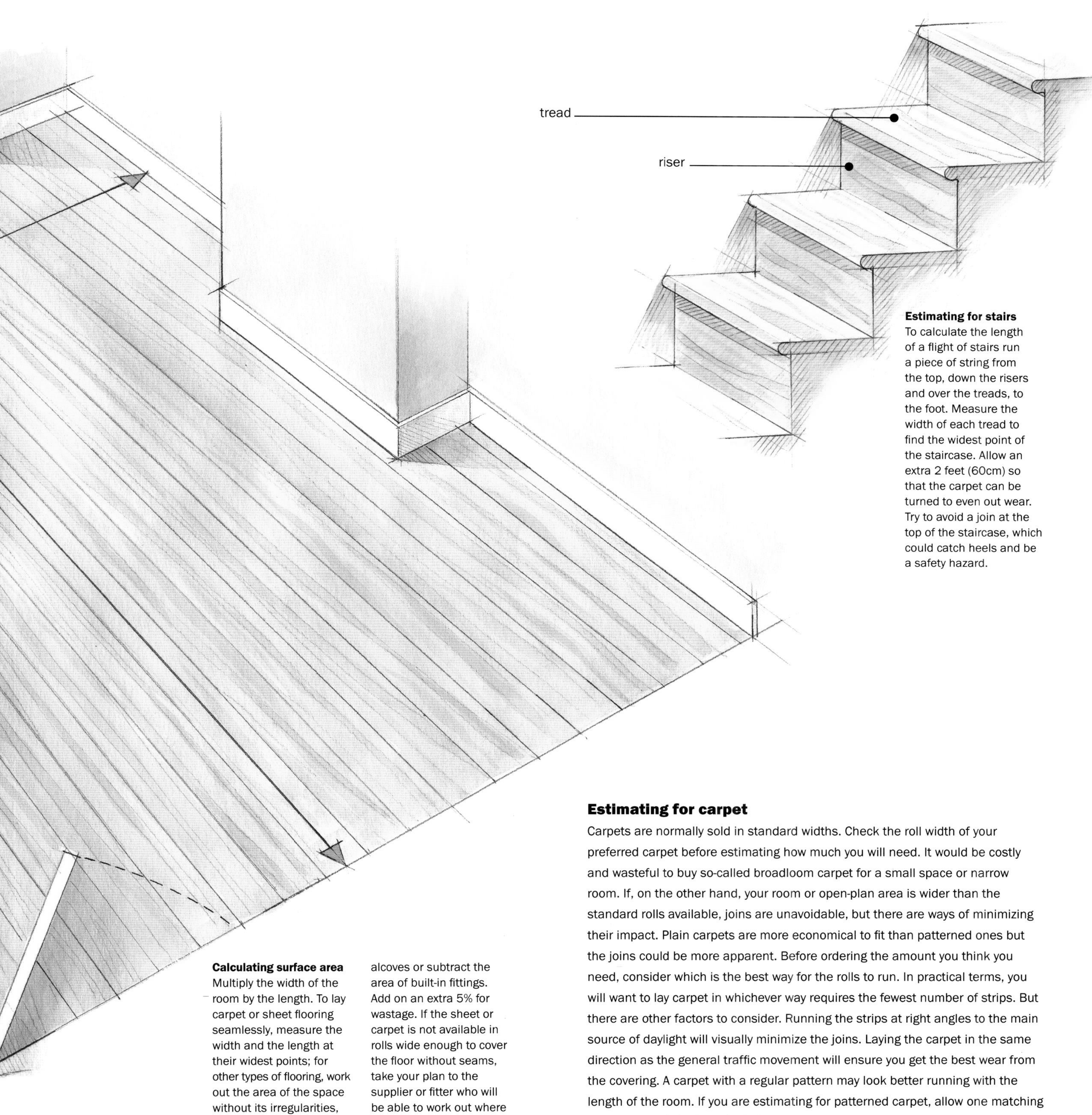

Estimating for stairs
To calculate the length of a flight of stairs run a piece of string from the top, down the risers and over the treads, to the foot. Measure the width of each tread to find the widest point of the staircase. Allow an extra 2 feet (60cm) so that the carpet can be turned to even out wear. Try to avoid a join at the top of the staircase, which could catch heels and be a safety hazard.

Calculating surface area
Multiply the width of the room by the length. To lay carpet or sheet flooring seamlessly, measure the width and the length at their widest points; for other types of flooring, work out the area of the space without its irregularities, then add in the area of alcoves or subtract the area of built-in fittings. Add on an extra 5% for wastage. If the sheet or carpet is not available in rolls wide enough to cover the floor without seams, take your plan to the supplier or fitter who will be able to work out where the seams should go.

Estimating for carpet

Carpets are normally sold in standard widths. Check the roll width of your preferred carpet before estimating how much you will need. It would be costly and wasteful to buy so-called broadloom carpet for a small space or narrow room. If, on the other hand, your room or open-plan area is wider than the standard rolls available, joins are unavoidable, but there are ways of minimizing their impact. Plain carpets are more economical to fit than patterned ones but the joins could be more apparent. Before ordering the amount you think you need, consider which is the best way for the rolls to run. In practical terms, you will want to lay carpet in whichever way requires the fewest number of strips. But there are other factors to consider. Running the strips at right angles to the main source of daylight will visually minimize the joins. Laying the carpet in the same direction as the general traffic movement will ensure you get the best wear from the covering. A carpet with a regular pattern may look better running with the length of the room. If you are estimating for patterned carpet, allow one matching repeat per strip in the calculations.

Preparing wood

Whatever finish you intend for a wooden floor, the surface must be thoroughly prepared. It is rare to find a floor that doesn't need years of buildup of dirt, old paint or polish removing, as well as any imperfections. Hiring an electrical sander is usually the most effective means of tackling the job.

Do not underestimate the scale of the task. You will need several days to complete the job – ensure you hire all equipment for a sufficient length of time. Most jobs require a drum floor sander, an edge sander and possibly a small disc sander, plus sheets of sandpaper to fit around the drum, in coarse, medium and fine grades. Machine sanding is physically strenuous, dusty and noisy. Warn neighbours about the row and if anyone in your family has respiratory problems, arrange for them to stay away from the house for a few days until the dust has cleared. For your personal protection you must wear a mask and goggles while sanding; ear protectors are also useful. Do read the operating and safety instructions that should be supplied to you by the hire shop.

Begin by clearing the room, covering what you can't remove with dust sheets. Seal off doorways with plastic sheeting and open windows. Cover computers and other sensitive electronic equipment elsewhere in the home.

Machine sanding

1 Try to keep the flex over your shoulder and fit the power plug into a circuit breaker. Keep the sander moving – don't dwell on one spot or it will gouge a hole. If boards are very uneven, dished or stained, sand diagonally using a coarse-grade sheet.

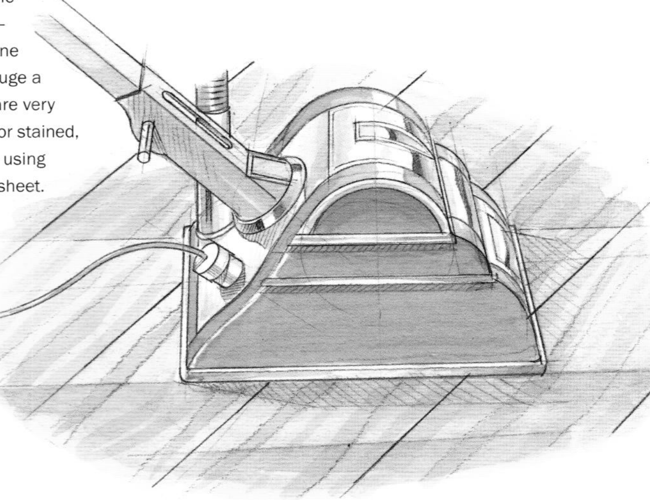

Preparing to sand

1 Whether or not the wood has been previously exposed, it is likely that some nails will have worked their way to the surface. Use a nail punch to sink the nail heads well below the surface and fill the holes with stopping.

2 Tackle old paint splashes with chemical stripper and a scraper, or sandpaper. If the floor has been painted previously, strip off as much paint as possible with chemical stripper, or the paint will clog up the sanding sheet. Wash the floor with hot water and detergent, but avoid overwetting it.

3 Check before sanding parquet, block or strip floors that they are either solid wood or that the veneer is thick enough to be sanded. If you are sanding herringbone parquet, work in both directions, following the pattern, using first medium- and then fine-grade sheets.

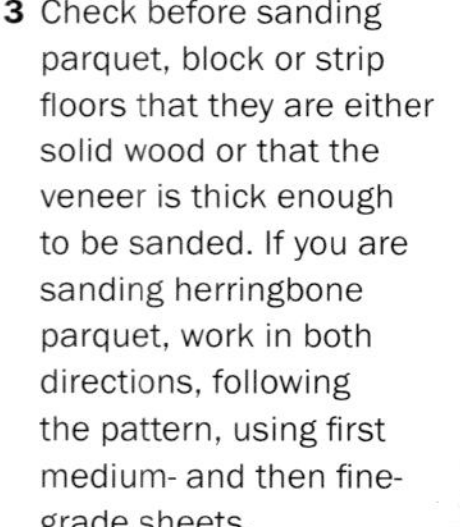

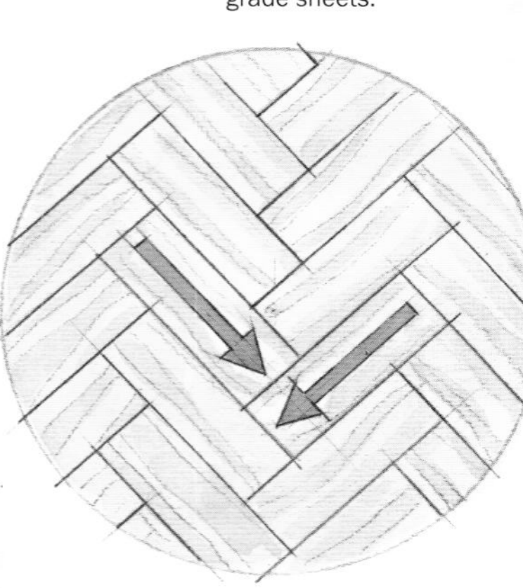

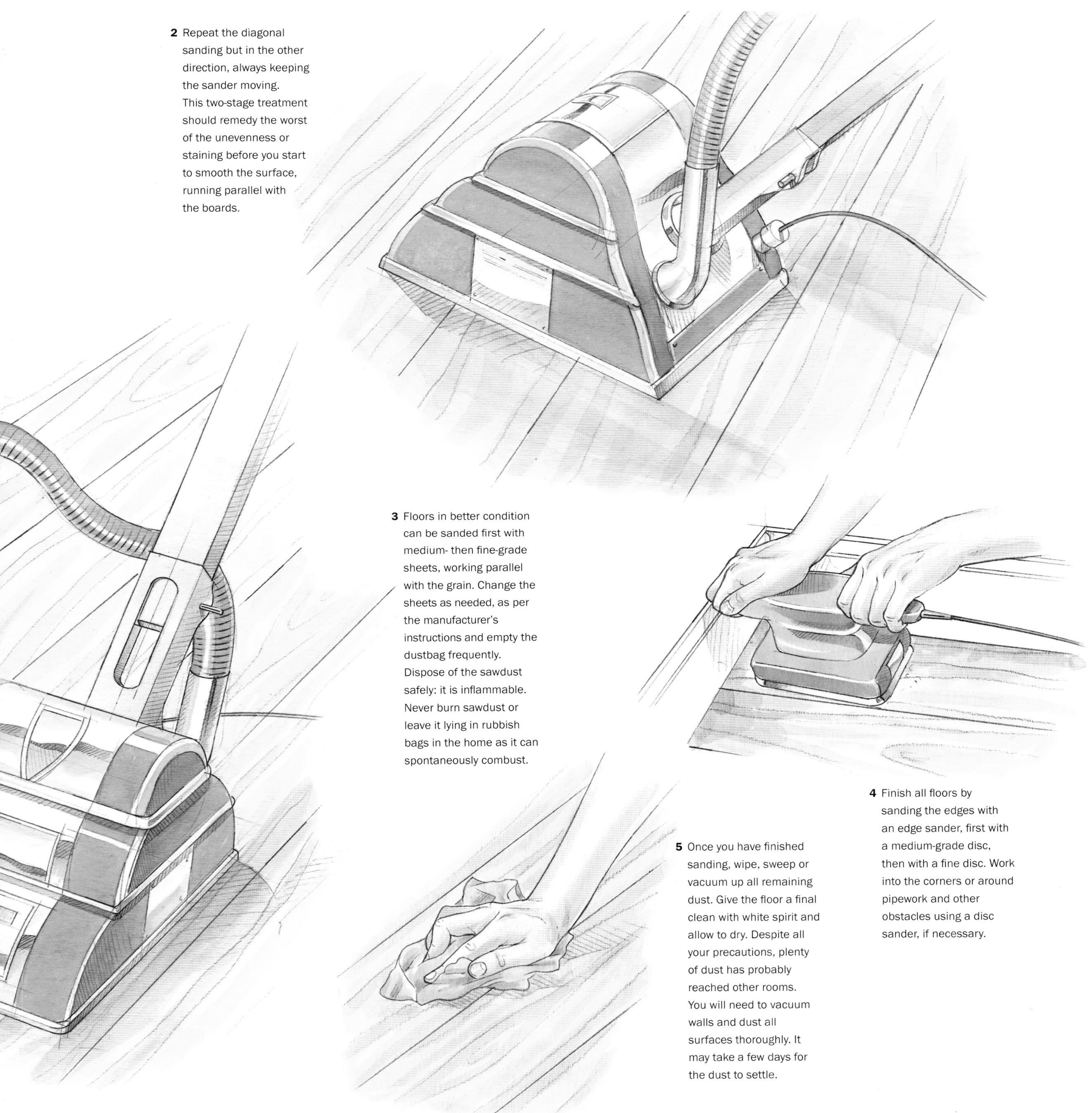

2 Repeat the diagonal sanding but in the other direction, always keeping the sander moving. This two-stage treatment should remedy the worst of the unevenness or staining before you start to smooth the surface, running parallel with the boards.

3 Floors in better condition can be sanded first with medium- then fine-grade sheets, working parallel with the grain. Change the sheets as needed, as per the manufacturer's instructions and empty the dustbag frequently. Dispose of the sawdust safely: it is inflammable. Never burn sawdust or leave it lying in rubbish bags in the home as it can spontaneously combust.

4 Finish all floors by sanding the edges with an edge sander, first with a medium-grade disc, then with a fine disc. Work into the corners or around pipework and other obstacles using a disc sander, if necessary.

5 Once you have finished sanding, wipe, sweep or vacuum up all remaining dust. Give the floor a final clean with white spirit and allow to dry. Despite all your precautions, plenty of dust has probably reached other rooms. You will need to vacuum walls and dust all surfaces thoroughly. It may take a few days for the dust to settle.

Decorative treatments for wood

A newly stripped wooden floor can lend itself to a number of treatments using stain and varnish or paint. Whatever treatment you choose, the bare floor must be free of dust or grease, and any knots treated with knotting solution. If you intend painting the floor, it must first be primed, unless you are opting for a limed finish. Staining needs no priming – the stain is intended to soak into the wood – but it does need good natural light to show it to its best effect. Make up swatches of the paint or stain you intend to use on bits of wood, with several coats of varnish on top, and look at them in the room to see how the light affects them.

Do not underestimate how long it takes to complete a decorated floor, especially an all-over treatment. You may need several coats of paint, for example, and each coat must be fully dry before the next can be applied. If you choose to use stain or dye, one coat may suffice, but it is essential to put on several coats of varnish over the stain to ensure a truly hard-wearing finish.

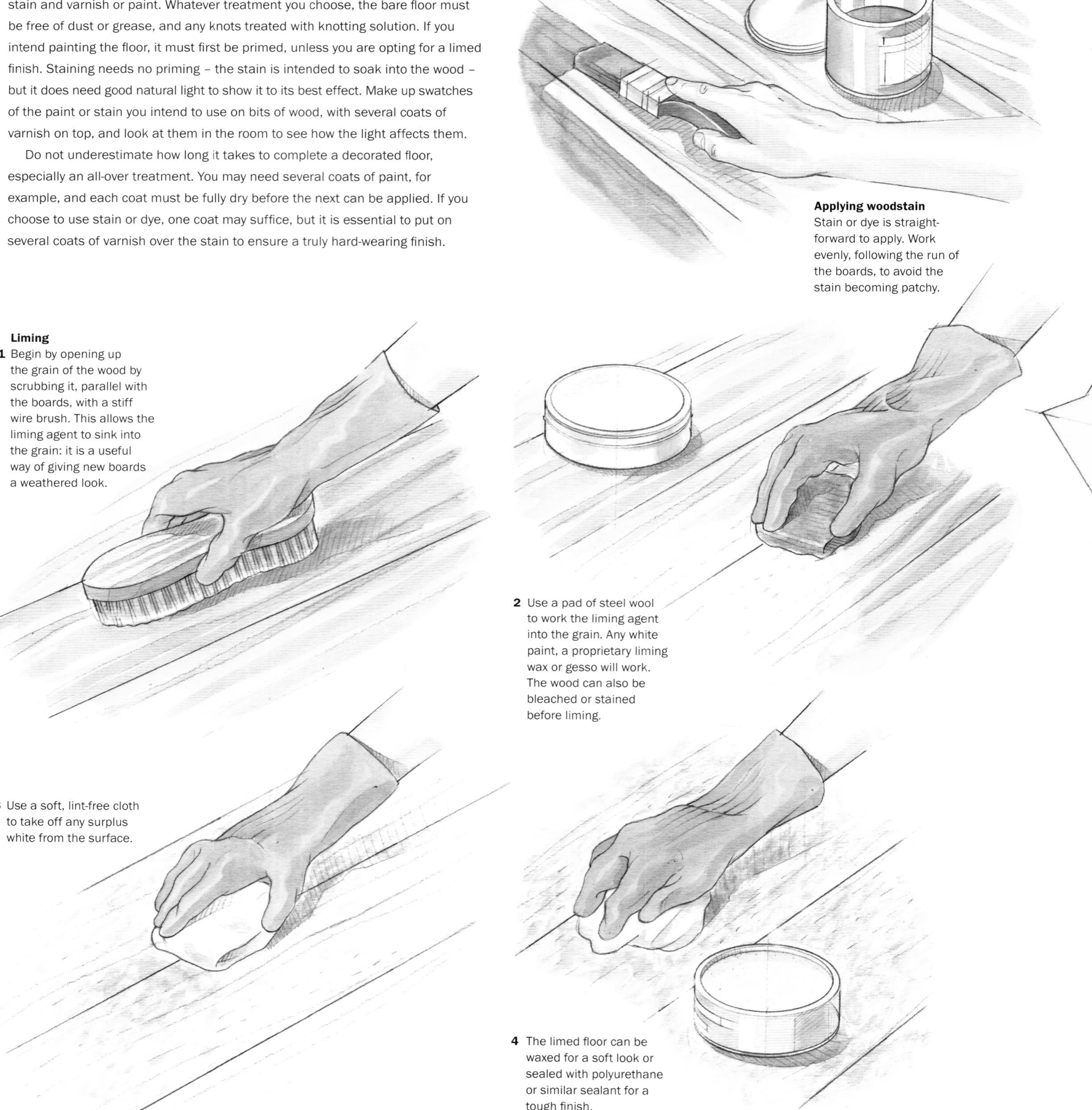

Applying woodstain
Stain or dye is straightforward to apply. Work evenly, following the run of the boards, to avoid the stain becoming patchy.

Liming

1 Begin by opening up the grain of the wood by scrubbing it, parallel with the boards, with a stiff wire brush. This allows the liming agent to sink into the grain: it is a useful way of giving new boards a weathered look.

2 Use a pad of steel wool to work the liming agent into the grain. Any white paint, a proprietary liming wax or gesso will work. The wood can also be bleached or stained before liming.

3 Use a soft, lint-free cloth to take off any surplus white from the surface.

4 The limed floor can be waxed for a soft look or sealed with polyurethane or similar sealant for a tough finish.

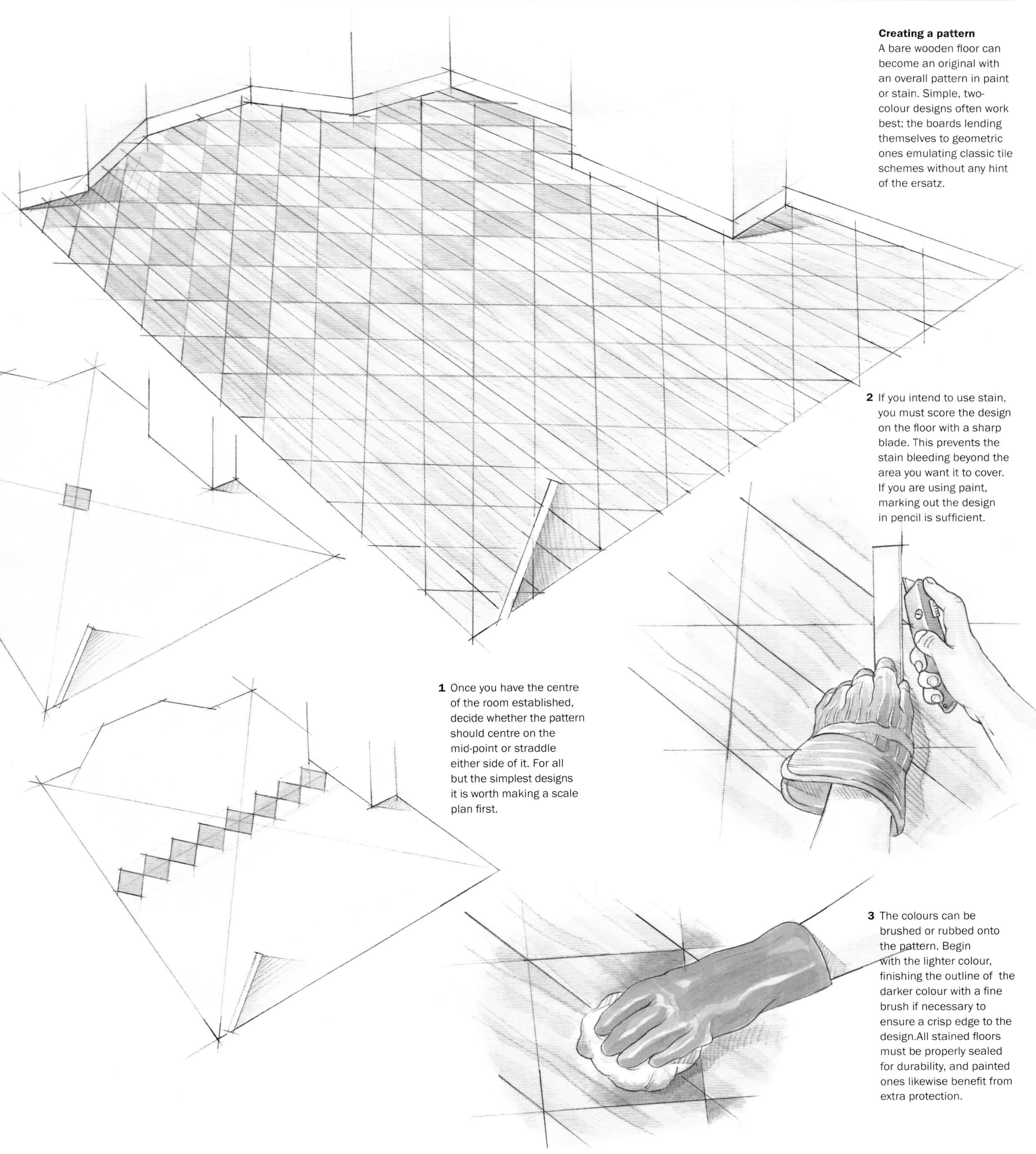

Creating a pattern

A bare wooden floor can become an original with an overall pattern in paint or stain. Simple, two-colour designs often work best; the boards lending themselves to geometric ones emulating classic tile schemes without any hint of the ersatz.

1 Once you have the centre of the room established, decide whether the pattern should centre on the mid-point or straddle either side of it. For all but the simplest designs it is worth making a scale plan first.

2 If you intend to use stain, you must score the design on the floor with a sharp blade. This prevents the stain bleeding beyond the area you want it to cover. If you are using paint, marking out the design in pencil is sufficient.

3 The colours can be brushed or rubbed onto the pattern. Begin with the lighter colour, finishing the outline of the darker colour with a fine brush if necessary to ensure a crisp edge to the design.All stained floors must be properly sealed for durability, and painted ones likewise benefit from extra protection.

Stencilling, spattering, combing

Stencilling a border gives a unified, finished look that works well in the right context. The hand-done effect is part of the charm: it can give an otherwise plain wood floor a quaint, personalized quality. Suitable paints include signwriter's colours, Japan colour, acrylics and poster colours and oil-based paints. Aerosol paints can be used but they are difficult to control and require large areas to be masked off. Subtle amounts of colour often work better than solid bold ones.

Spattering and combing effects represent a complete departure from the grain and tones of the wood, and some interesting trompe l'oeil effects can easily be achieved. Spattering techniques range from flecking one or more colours of paint from a stiff brush to produce a random speckled effect, to a more 'mineralized' look produced by flecking white spirit or water over wet paint which produces a pleasing, pebble-like finish. Combing, as the term implies, is paint applied with a wide-toothed comb following the grain of the wood, or in squares or randomly.

With all these treatments, it is vital to practise on paper or offcuts of wood before tackling the floor itself to gain confidence and try out different colours and effects. All paint effects must be coated with several coats of transparent seal.

Planning a border

1 If the stencil has a symmetry, begin working from the centre of a focal point, such as a bay or chimney breast, on the wall where the design is going to be most obvious. Ending behind the door is the best way of adjusting the pattern if it doesn't quite finish evenly.

2 Whether you make or buy a stencil, do make sure it is made of good quality material. Choose a pattern that is unfussy. Its size and design should help you decide how far from the room's perimeter to paint the border.

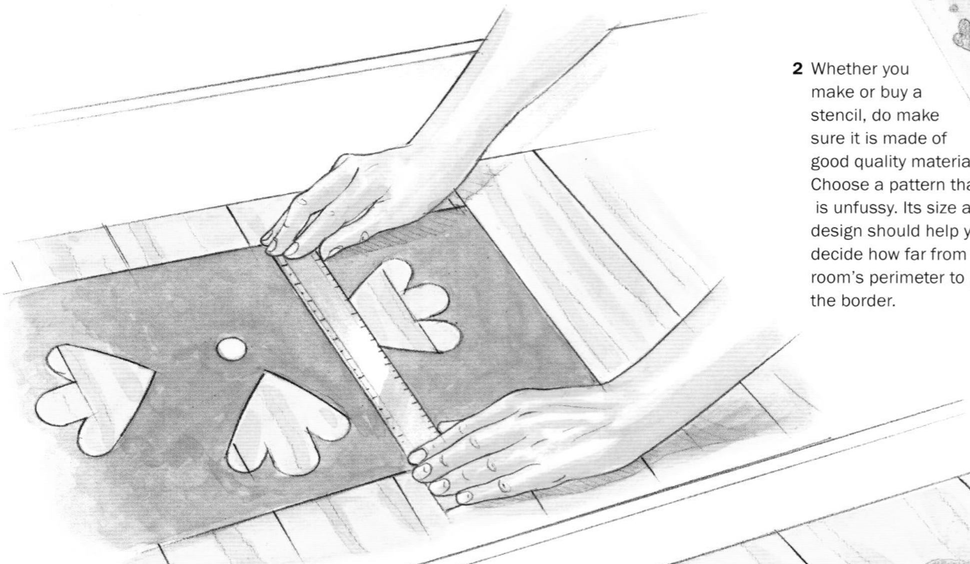

3 Ensure there is enough solid space between the cutouts so that the design remains distinct as you paint through the holes. Keep the position of the stencil equidistant from the last: the floorboards themselves may help in this respect. The final effect will look more planned.

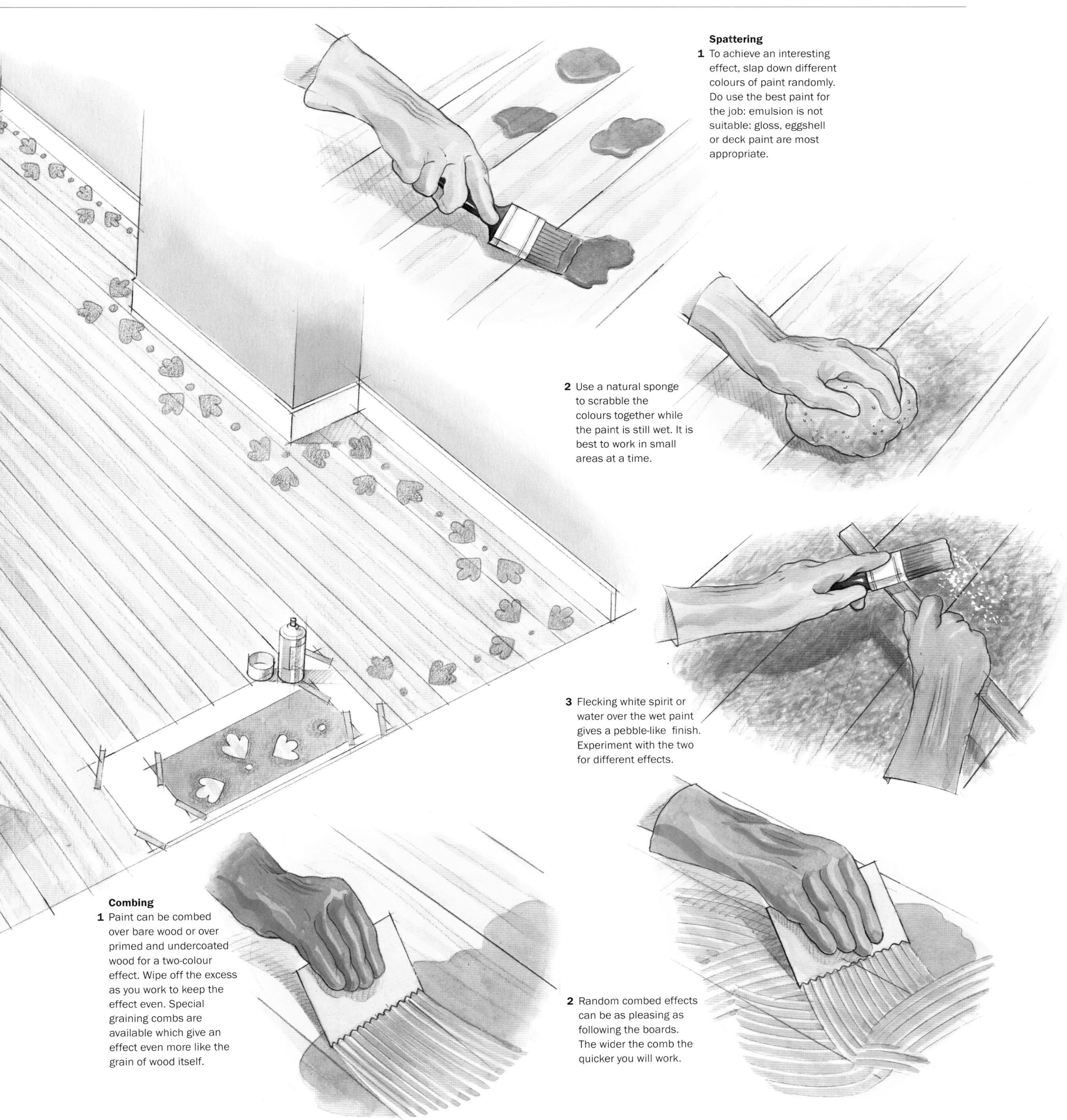

Spattering

1 To achieve an interesting effect, slap down different colours of paint randomly. Do use the best paint for the job: emulsion is not suitable: gloss, eggshell or deck paint are most appropriate.

2 Use a natural sponge to scrabble the colours together while the paint is still wet. It is best to work in small areas at a time.

3 Flecking white spirit or water over the wet paint gives a pebble-like finish. Experiment with the two for different effects.

Combing

1 Paint can be combed over bare wood or over primed and undercoated wood for a two-colour effect. Wipe off the excess as you work to keep the effect even. Special graining combs are available which give an effect even more like the grain of wood itself.

2 Random combed effects can be as pleasing as following the boards. The wider the comb the quicker you will work.

Laying woodblock and woodstrip

Laying a new wooden floor is a fairly straightforward project. Manufactured woodblocks and strips, either tongued and grooved or straight edged, are widely available and can be laid directly onto a solid or suspended floor. A concrete floor needs to be damp-proofed and overlaid with screed. The condition of existing floorboards will affect your choice: if they are sound, you can lay woodblock directly on top, either fixed by 'secret nailing', by driving panel pins diagonally through the side of the blocks into the floor below, or, far simpler, by gluing the blocks, which cuts down on creaks. Many are sold tongued-and-grooved, which can be laid as a 'floating floor', where only the perimeter blocks need be fixed to the underlying surface. If the existing boards are not sound, you will need to take them up completely and opt for woodstrip, which is laid directly over the joists.

Laying woodblock

1 If you are covering old floorboards, plane off any unevenness before laying new blocks.

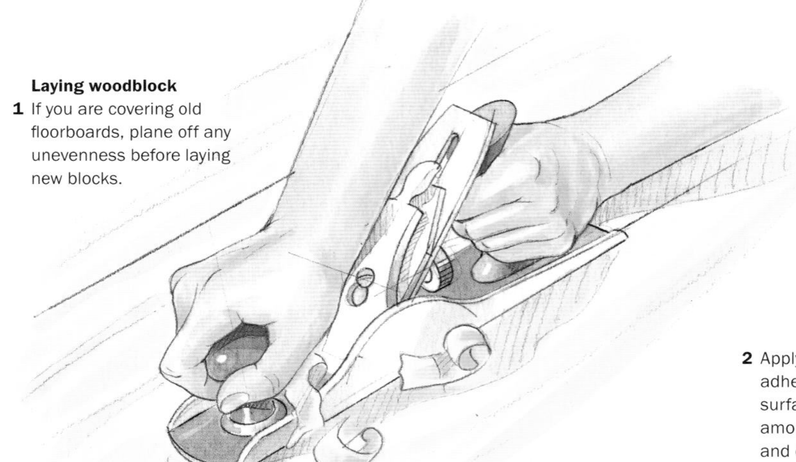

2 Apply a recommended adhesive to the underlying surface. Try to keep the amount you apply even, and only cover a small area at a time.

3 Start laying the new blocks from one corner, at right angles to the underlying boards. Plan the pattern the blocks create from the outset: staggering the joins, like courses of bricks, is the most stable and also aesthetically pleasing.

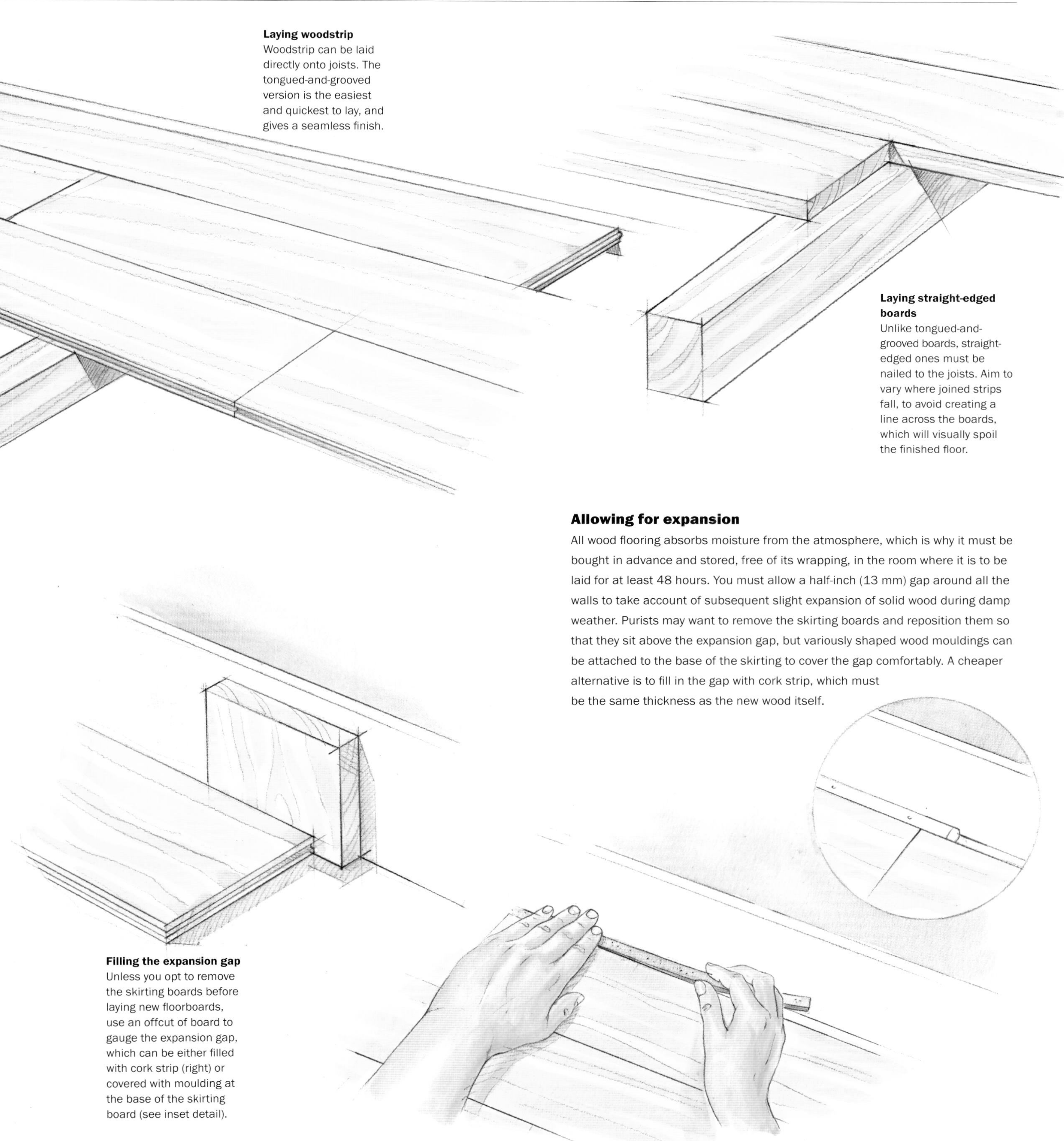

Laying woodstrip
Woodstrip can be laid directly onto joists. The tongued-and-grooved version is the easiest and quickest to lay, and gives a seamless finish.

Laying straight-edged boards
Unlike tongued-and-grooved boards, straight-edged ones must be nailed to the joists. Aim to vary where joined strips fall, to avoid creating a line across the boards, which will visually spoil the finished floor.

Allowing for expansion

All wood flooring absorbs moisture from the atmosphere, which is why it must be bought in advance and stored, free of its wrapping, in the room where it is to be laid for at least 48 hours. You must allow a half-inch (13 mm) gap around all the walls to take account of subsequent slight expansion of solid wood during damp weather. Purists may want to remove the skirting boards and reposition them so that they sit above the expansion gap, but variously shaped wood mouldings can be attached to the base of the skirting to cover the gap comfortably. A cheaper alternative is to fill in the gap with cork strip, which must be the same thickness as the new wood itself.

Filling the expansion gap
Unless you opt to remove the skirting boards before laying new floorboards, use an offcut of board to gauge the expansion gap, which can be either filled with cork strip (right) or covered with moulding at the base of the skirting board (see inset detail).

Laying soft tiles and sheet flooring

Soft tiles and sheet flooring materials are reasonably straightforward to lay. The technique for laying vinyl, linoleum, cork, rubber and carpet tiles is the same, the only difference may be which adhesive to use, but many are self-adhesive. The format of tiles makes them easier to fit than sheeting, but it is important to spend some time deciding how best to lay them, usually working out from the centre of the room, though not necessarily the true centre. For a neat and professional finish the trick is to avoid having an overly narrow border of cut tiles around the perimeter, which looks mean and awkward.

Large rolls of sheet flooring, such as vinyl, are more difficult to handle. Get someone to help you at this stage. Decide whether it would be easier to remove doors first, to get the sheet down with ease. You'll need to cut it roughly to size (allowing a trimming margin of at least 6 inches (15cm) on all sides), then make cuts at the corners so that it can be laid flat. Then stand back to judge whether the sheet is straight, especially if it has a pattern.

Cushioned vinyl and some 'stay-flat' sheets do not need to be stuck down, other than at the edges and along any seams. For both tiles and sheet flooring, it may be worth hiring a floor roller to ensure a perfectly flat finish.

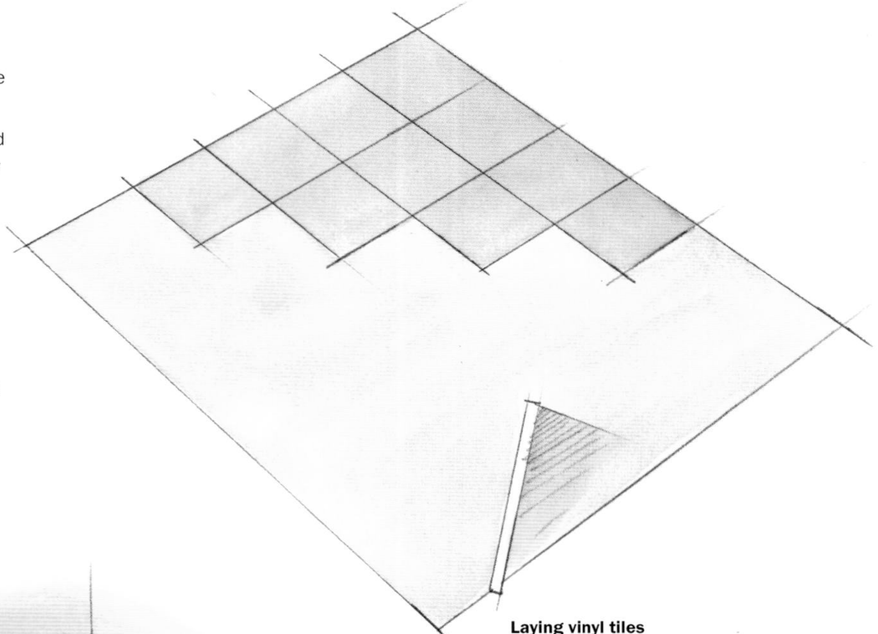

Laying vinyl tiles

1 To avoid an awkward border, calculate the centre of the room (see page 166), and loose lay tiles in one of the four quarters, working away from the centre point.

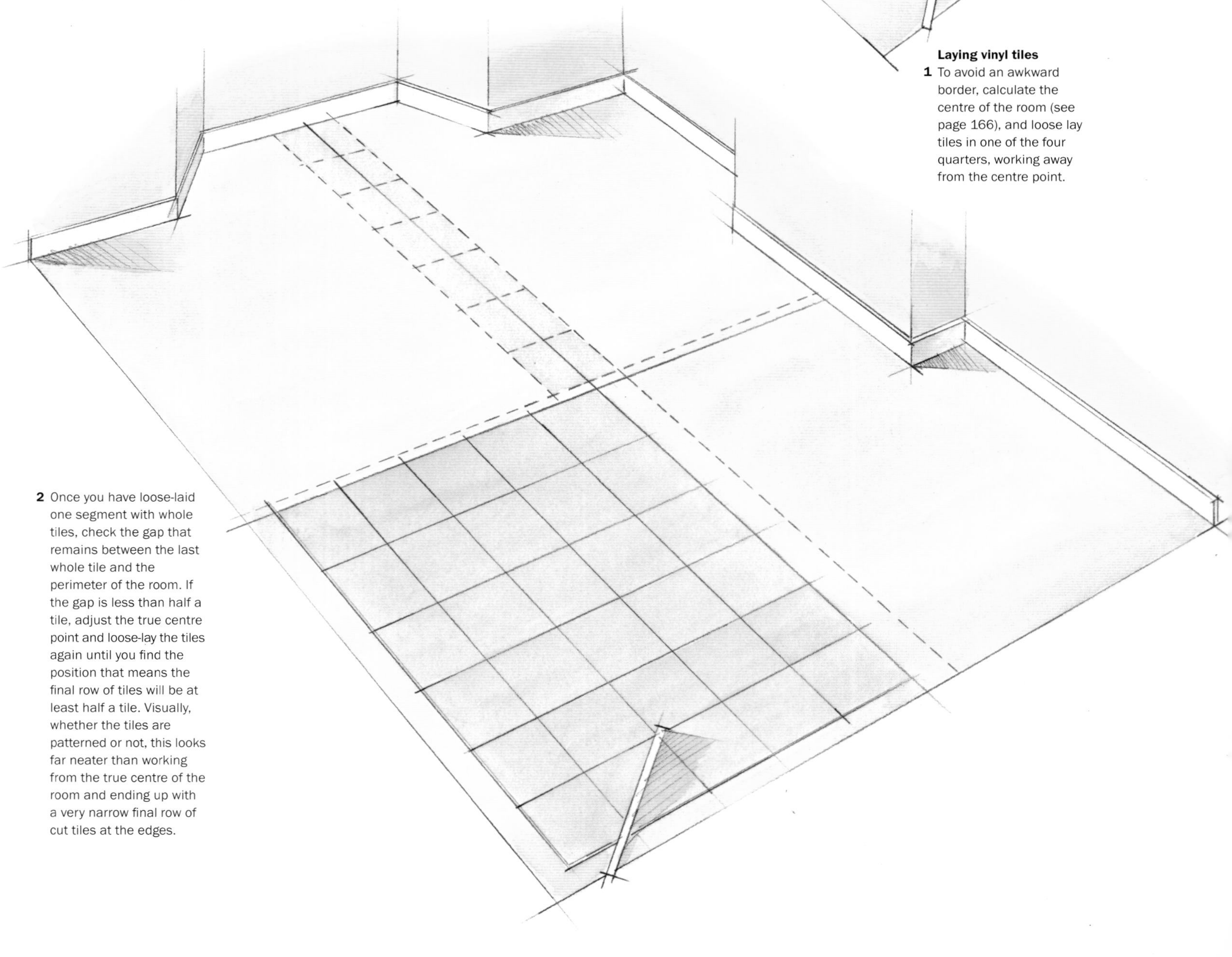

2 Once you have loose-laid one segment with whole tiles, check the gap that remains between the last whole tile and the perimeter of the room. If the gap is less than half a tile, adjust the true centre point and loose-lay the tiles again until you find the position that means the final row of tiles will be at least half a tile. Visually, whether the tiles are patterned or not, this looks far neater than working from the true centre of the room and ending up with a very narrow final row of cut tiles at the edges.

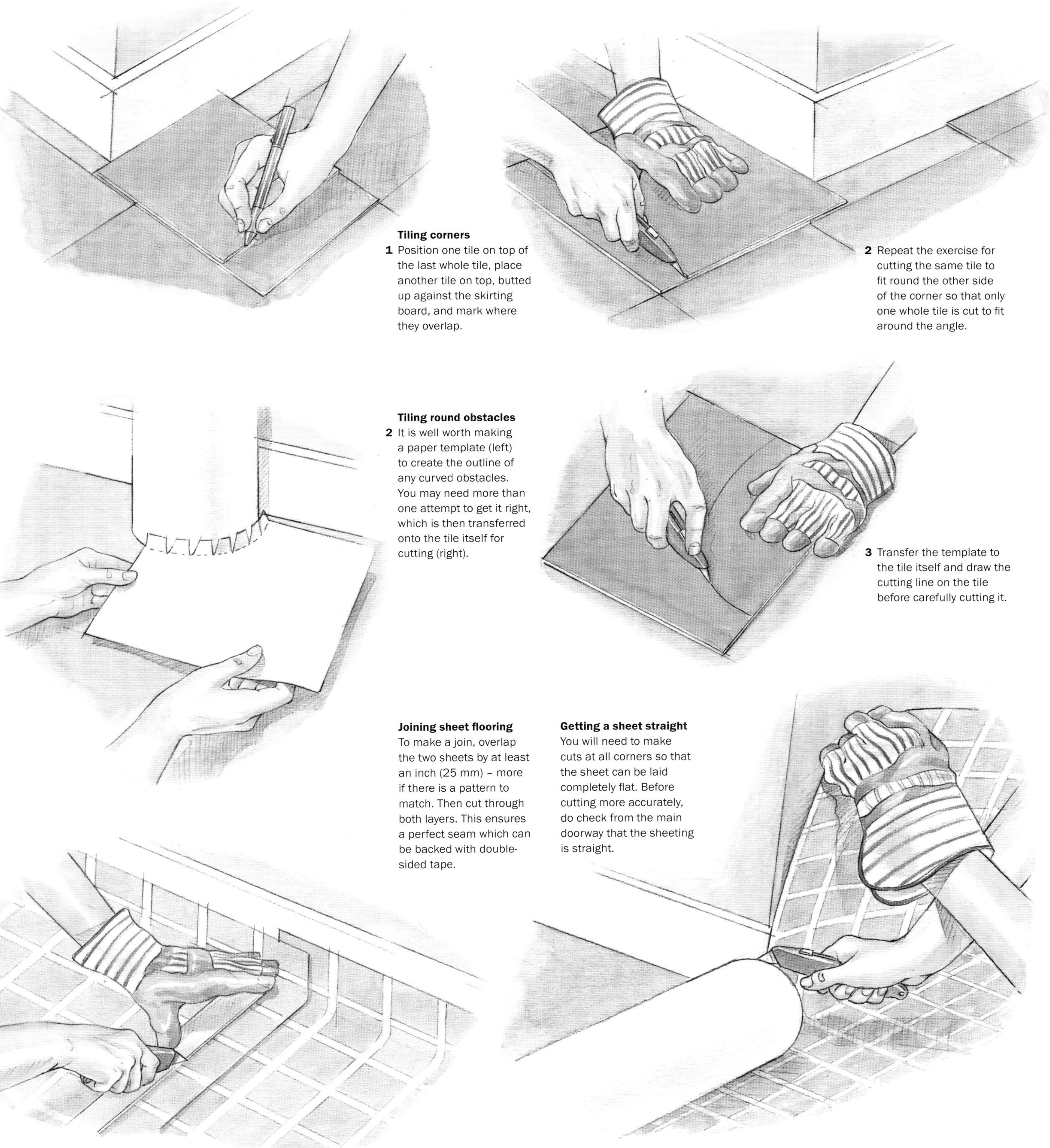

Tiling corners

1 Position one tile on top of the last whole tile, place another tile on top, butted up against the skirting board, and mark where they overlap.

2 Repeat the exercise for cutting the same tile to fit round the other side of the corner so that only one whole tile is cut to fit around the angle.

Tiling round obstacles

2 It is well worth making a paper template (left) to create the outline of any curved obstacles. You may need more than one attempt to get it right, which is then transferred onto the tile itself for cutting (right).

3 Transfer the template to the tile itself and draw the cutting line on the tile before carefully cutting it.

Joining sheet flooring

To make a join, overlap the two sheets by at least an inch (25 mm) – more if there is a pattern to match. Then cut through both layers. This ensures a perfect seam which can be backed with double-sided tape.

Getting a sheet straight

You will need to make cuts at all corners so that the sheet can be laid completely flat. Before cutting more accurately, do check from the main doorway that the sheeting is straight.

Laying foam-backed carpet

Laying foam-backed carpet is within the scope of most DIY enthusiasts because unlike fabric-backed carpet it does not have to be stretched or gripped. Nor does it need a separate underlay. Foam-backed carpet can be stuck down or, for a cheaper, quicker result, stapled or tacked, although this does not give such a neat finish. If you are laying over floorboards, put down a paper underlay, which prevents dirt blowing up from between the boards and accumulating under the carpet where it could cause damage. Hardboard over uneven boards makes a good smooth surface. Concrete floors, provided they are perfectly dry, need no such underlay.

Unless you are confident, start with a small room before tackling a large area. If there are obstacles, say a washbasin pedestal, it is best to make a template by covering the floor with paper, leaving a small margin, and using a pen held against a small block to scribe around the perimeter. For less awkward jobs or large surfaces it is sufficient to cut the carpet roughly to size, allowing say 6 inches (150mm) for trimming on all sides. Keep some of the waste: it may be useful for patching later.

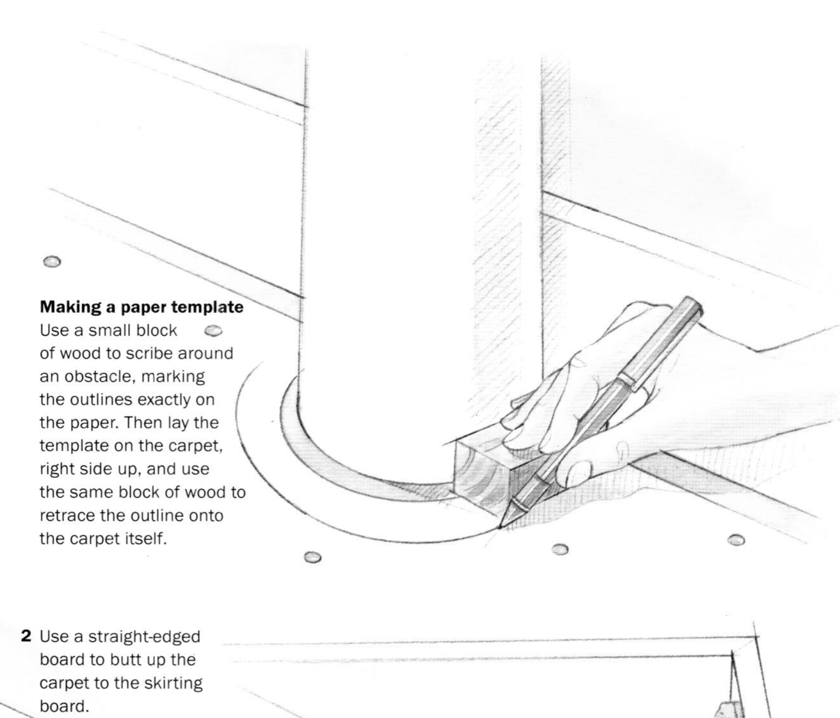

Making a paper template
Use a small block of wood to scribe around an obstacle, marking the outlines exactly on the paper. Then lay the template on the carpet, right side up, and use the same block of wood to retrace the outline onto the carpet itself.

Laying the carpet

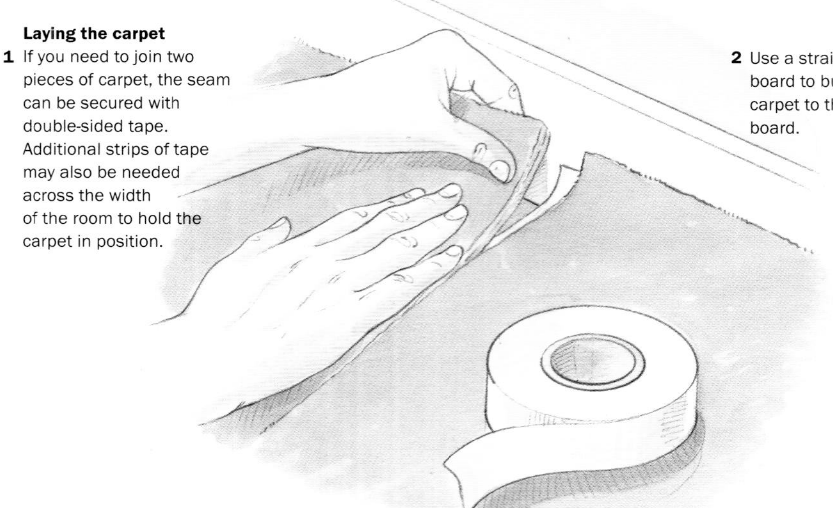

1 If you need to join two pieces of carpet, the seam can be secured with double-sided tape. Additional strips of tape may also be needed across the width of the room to hold the carpet in position.

2 Use a straight-edged board to butt up the carpet to the skirting board.

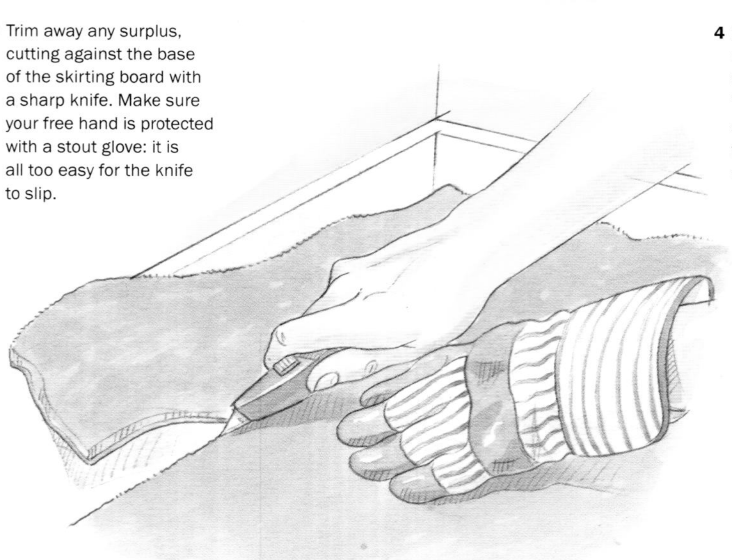

3 Trim away any surplus, cutting against the base of the skirting board with a sharp knife. Make sure your free hand is protected with a stout glove: it is all too easy for the knife to slip.

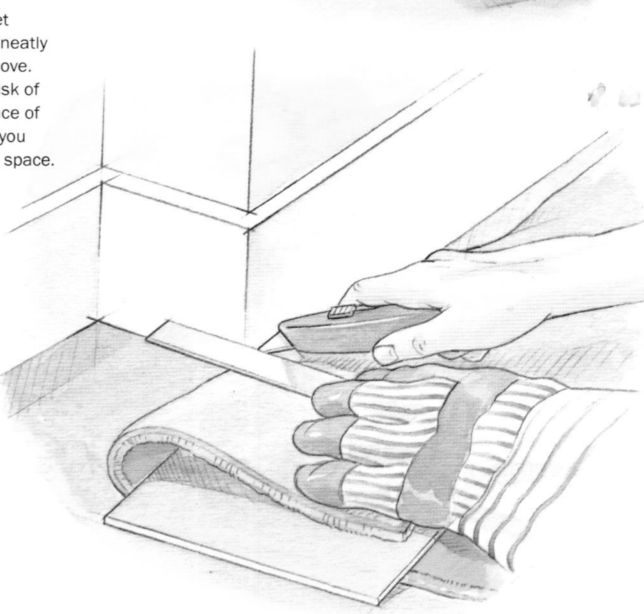

4 Fold back the carpet over a board to cut neatly to fit a corner or alcove. This will avoid the risk of damaging the surface of the carpet itself as you work in an awkward space.

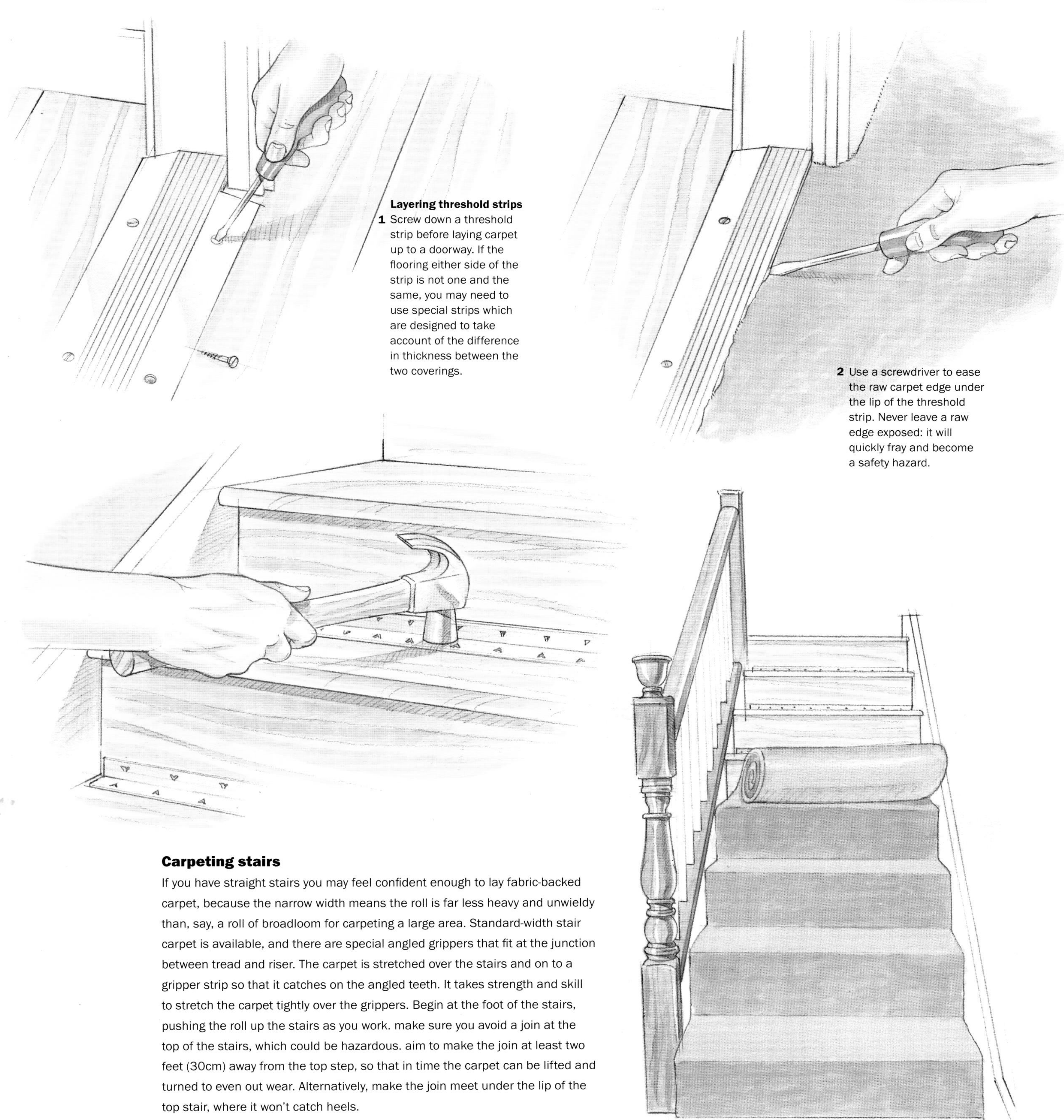

Layering threshold strips

1 Screw down a threshold strip before laying carpet up to a doorway. If the flooring either side of the strip is not one and the same, you may need to use special strips which are designed to take account of the difference in thickness between the two coverings.

2 Use a screwdriver to ease the raw carpet edge under the lip of the threshold strip. Never leave a raw edge exposed: it will quickly fray and become a safety hazard.

Carpeting stairs

If you have straight stairs you may feel confident enough to lay fabric-backed carpet, because the narrow width means the roll is far less heavy and unwieldy than, say, a roll of broadloom for carpeting a large area. Standard-width stair carpet is available, and there are special angled grippers that fit at the junction between tread and riser. The carpet is stretched over the stairs and on to a gripper strip so that it catches on the angled teeth. It takes strength and skill to stretch the carpet tightly over the grippers. Begin at the foot of the stairs, pushing the roll up the stairs as you work. make sure you avoid a join at the top of the stairs, which could be hazardous. aim to make the join at least two feet (30cm) away from the top step, so that in time the carpet can be lifted and turned to even out wear. Alternatively, make the join meet under the lip of the top stair, where it won't catch heels.

Maintenance

Your decision when choosing a flooring material should in part be influenced by the maintenance required once a floor is laid. You need to have an idea of how much effort (and possible expense) you are prepared to devote to its care and whether your choice is sensible, given the location for which it is intended. Different materials require cleaning in different ways, and you should always use products and techniques recommended by the suppliers or manufacturers to ensure a long life for the floor. This final section gives an indication of the ease – or otherwise – of maintaining all the flooring materials discussed in the book.

Hard floors

Brick

Make sure any mortar residue is cleaned from a new brick floor immediately after laying to prevent staining. You may need to use a proprietary cleanser. Once laid, though, bricks generally need little in the way of after-care. They can be polished, dressed with linseed oil or sealed – sparingly (not recommended at all for absorbent bricks) but none of these treatments is necessary and can increase the slipperiness of the surface. You will need to remove the old polish completely from time to time to prevent a build up of dirt. General maintenance consists of sweeping regularly and washing with a mild detergent and rinsing thoroughly.

Stone

Most limestones and sandstones are porous and stain readily. Sealing will prevent staining but it is not always recommended, and it can alter the appearance of the floor. Never polish smooth-textured stone, which will only make it dangerously slippery.

To clean stone floors, wash with warm water and a non-caustic, sulphate-free detergent, then rinse. Never use alkaline soaps on marble which will corrode the surface irreparably. Specialist cleaning may be required for stubborn stains. Antique stone reclaimed from old houses may need to be steam-pressure cleaned.

Hard tiles

Porous tiles (terracotta, quarry and encaustic) need to be sealed to inhibit staining with a proprietary sealer or a dressing of two coats of beeswax mixed with boiled linseed oil. Sealing needs to be carried out before the tiles are fixed and after grouting. The seal must be applied to completely clean and dry tiles, otherwise moisture and dirt will be trapped in the tiles. Apply a light even coat of seal so that it doesn't dry in streaks or blotches.

General maintenance of hard tiles is simple and straightforward: wash with warm water and a mild detergent and rinse thoroughly. Any white patches appearing on porous tiles are the result of salt deposits in the tiles: further washing with clean water should suffice to remove these patches. A wax paste applied to thoroughly dry tiles gives a final finish. Apply two or three coats after sealing then weekly for the first six weeks to create a hard-wearing surface.

Mop up spills as soon as possible using a soft cloth and clean water. Acidic liquids, including citrus juices, Coca-Cola, wine, spirits, vinegar and urine are liable to cause stains, especially on porous titles. White spirit may lift stubborn marks.

Glazed tiles need no sealing. Wash with a mild detergent and rinse. If you wish to polish them, use only small amounts, renewed from time to time. Dirty grouting can be cleaned with a proprietary tile bleach.

To restore an old encaustic floor, use a floor scrubber to lift out the dirt, and wire wool to work gently into the corners. Encaustic tiles tend to stain easily before they are sealed and need to be handled with care.

Terrazzo and mosaic

The smooth surface of terrazzo is fairly non-slip, except when wet, washed with soap or polished. General maintenance is simple: wash with a little scouring powder and rinse. Avoid using a polish which contains wax as this will make the surface too slippery.

Mosaic needs washing with mild detergent. Never apply a high polish or the natural key provided by the difference in texture between mosaic and grouting will be lost and the surface will be dangerously slippery.

Concrete, metal and glass

Concrete floors must be sealed to prevent the surface from dusting. Once sealed, maintain by scrubbing with hot water and detergent. Metal and glass are cleaned in the same way, too. Do ensure that any glass surfaces are thoroughly dry before walking over them.

Wood

Properly sealed (see page 100) and finished wood is straightforward to keep clean on a daily basis. Wax polishes will need to be renewed every couple of months but most seals will last far longer and can be patch repaired in area of heavy traffic. Vacuum or sweep up loose dirt and wipe over the surface with a damp cloth or mop. Avoid overwetting the floor.

Despite being fairly low maintenance on a regular basis, timber floors demand more in the long term. Once the seal has begun to wear all over it must be replaced before dirt and moisture can penetrate the wood and begin to degrade it. Really worn floors may need refinishing (see pages 168–69).

Avoid stiletto heels at all costs, even at the risk of embarrassing your guests!. There is no quicker or more effective way of ruining any kind of timber floor than to subject it to the type of intense point load exerted by spike heels. The pockmarks not only break the seal but leave lasting dents that can only be eradicated by deep sanding (which would be impossible on a wood veneer floor).

Sheet and soft tiling

Cork

Cork needs to be sealed to protect it from soiling and to make it water-resistant. Presealed cork can be given an extra coat of seal after laying to prevent moisture from penetrating joints. Alternatively, unsealed cork can be wax-polished or dressed with organic primers. Cork flooring with a vinyl finish is much more resistant and requires no subsequent treatment. Seals and polishes should be renewed when they show signs of wear. Cork can be lightly sanded if it has become pitted or soiled. It will gradually fade a little in strong sunlight. Natural cork is very susceptible to cigarette burns and alkaline chemicals.

A cork floor demands dedicated upkeep to preserve its appearance. It is important to make sure the cork is kept clean and free from grit which might break down the seal and allow dirt and moisture through. Sweep or vacuum regularly, wipe with a damp cloth, and polish occasionally and sparingly. Spills must be mopped up immediately.

Rubber

Natural rubber is readily marked by fats and solvents, but even synthetic rubber can be damaged if left unpolished. Use polish sparingly and renew it from time to time.

Clean the floor regularly with a mild detergent. Go over the surface with a damp mop rinsed out in clear water to remove all traces of detergent, which otherwise might leave a residue. Take care with a rubber floor which has a smooth finish: the surface can be very slippery when wet.

Linoleum

Linoleum needs no sealing but buffing it with an emulsion polish can bring up a glossy sheen. Before polishing, allow up to 48 hours after laying to give the adhesive time to cure.

For regular maintenance, dust mop or vacuum to remove loose grit and dirt which can scratch the surface and use a damp mop with a mild detergent on soiled floors. Avoid overwetting the floor. Strip off the polish occasionally with a recommended stripper and renew. Mop up spills as soon as they occur. Linoleum resists most stains, but solvents such as petrol, dry cleaning fluid, nail varnish and nail varnish remover, washing soda and oven cleaner can do lasting damage. Cigarette burns can be rubbed or buffed away.

Vinyl

Vinyl is waterproof and resistant to oils, fats and most household chemicals, which is why it is such an easy-care, all-purpose flooring for hardworking areas of the home, although smooth vinyl is slippery when wet.

Unlike natural materials, which may mellow with time, worn vinyl is merely shoddy. For this reason, you must ensure that the right cleaning products are used and that the floor is protected from the type of spills most likely to damage it, as well as from grit. Black rubber heel marks can be particularly detrimental; if not tackled immediately, anti-oxidants in the rubber can stain permanently. Rubber-backed rugs and mats can cause similar problems. Vinyl is badly damaged by cigarette burns. Premature damage can also result from using the wrong cleaners, such as bleach, scouring powder, liquid-based abrasives and strong alkaline detergents. Permanent damage or stains can be caused by stiletto heels, hot objects, cigarette burns, road tar, juice cordials or syrups, petrol, paraffin, white spirit, bleach, dry cleaning fluid, nail varnish remover, paint stripper, acids generally, and hair preparations.

Cushioned vinyl is particularly likely to be damaged by moving heavy objects, such as household appliances, over the floor. To protect the surface, place a piece of carpet underneath the appliance so that it can be pushed or pulled across the floor; alternatively 'walk' the appliance over a sheet of hardboard or plywood.

For general maintenance, wash with a damp mop and neutral or mild detergent. Rinse thoroughly to avoid a build-up of detergent which might dull the surface. Avoid overwetting the floor.

Polish with a water-wax emulsion polish. Do not use polish which contain solvents. Sweep or vacuum regularly, because any grit will abrade the surface of the material and cause it to wear through. If vinyl is laid in hallways or entrances, make sure you provide a doormat to remove loose dirt and debris, but choose one that is not latex or rubber-backed as the rubber can stain.

Leather

To protect and enrich leather, a newly laid floor should be waxed and buffed at least three times before you walk on it. Subsequently, buff the floor every two weeks and wax it about twice a year.

Waxing provides basic moisture-resistance and spills, if tackled immediately, will wipe off. Scratches are inevitable, but add to the character; a well-used floor has great depth of appeal.

Soft floors

Carpet

New carpet sheds a great deal of fluff in the first few weeks. During this period, it is best not to vacuum or to subject the floor to heavy wear. When the carpet has settled down, vacuum at least once a week. In areas where traffic is heaviest, more frequent vacuuming may be necessary. Animal hairs can be particularly tenacious: work slowly over the area, using a stiff brush as well if necessary.

Like any textile, carpet does stain. Stain-inhibition treatments, which may be either integral or applied after the carpet is laid, will only minimize damage, not entirely prevent it. Tackle spills immediately. Blot liquids with a damp cloth or paper towel to get as much off the surface as possible before the spill has a chance to be absorbed into the pile. Scrape up more solid deposits with a blunt knife. Do not add any additional liquid or you will spread the stain. After you have cleared or blotted as much as you can, wash the area gently, working from the outer rim of the stain to the centre. Do not scrub or you may damage the fibres and force the stain down into the pile. Avoid over-wetting at all costs.

Stain treatment depends on whether the stain is water- or oil-based. For water-based stains use a gentle cleaning solution made from warm water and a dash of mild liquid detergent (such as a detergent for hand-washing wool and silk, or carpet shampoo). For oil-based stains use dry cleaning fluid or solvent. Damp a cloth with the cleaning solution and dab the stain, working from the outside in. Blot to dry.

Stain checklist:

Alcohol —Blot and wash with detergent. Never apply salt to wine stains
Blood —Blot and wash with a cold-water detergent solution
Burn —Cut off burned fibres and wash with detergent
Chewing gum —Freeze with ice wrapped in plastic, scrape off and clean with dry cleaning fluid
Coffee— Blot and wash with detergent. When dry, use dry cleaning fluid to take out the grease from milk or cream.
Egg —Scrape off and clean with dry cleaning fluid
Fats and oils —Blot or scrape, as appropriate, and clean with dry cleaning fluid
Grass —Clean with dry cleaning fluid
Ink —Blot and wash with detergent or clean with dry cleaning fluid, depending on type
Juice— Blot and wash with detergent
Milk —Blot and wash with detergent. When dry, clean with dry cleaning fluid
Mud —Allow to dry, then vacuum or brush up dirt. Wash with detergent
Paint— Scrape and then wash with detergent (water-based paint) or clean with dry cleaning fluid (oil-based paint)
Shoe polish— Scrape and clean with dry cleaning fluid
Tar —Scrape and clean with dry cleaning fluid
Tea —Blot and wash with detergent. When dry, clean with dry cleaning fluid to remove grease from milk.
Urine —Blot, wash with detergent with a dash of antiseptic
Vomit —Scrape, wash with detergent with a dash of antiseptic
Wax —Scrape, applying mild heat to soften, then clean with dry cleaning fluid

Once a year carpets may be shampooed using a proprietary dry foam shampoo: follow the instructions and do not overwet. For a more thorough overhaul, it may be worth calling in professional carpet cleaners. Such firms use 'wet' vacuum cleaners to spray jets of cleaning solution into the pile. Carpets can take up to a day to dry after this treatment.

Natural fibre coverings

Many natural fibres are susceptible to wear from castors, so it is best to have additional protection under chair and sofa legs. High, spiky heels are also best avoided. Bright sunshine may cause fading in some varieties. Many of them, with the exception of rush react differently and usually adversely to humidity. Coir expands with moisture and may wrinkle and buckle. As it dries out, joints may open up. Sisal, on the other hand, shrinks when wetted. In very damp conditions all natural fibres rot.

For regular maintenance, all of these floorings should be vacuumed. Never wash or shampoo. Coir, sisal and jute are best treated with a stain-inhibition treatment. Tackle spills immediately, blotting up as much as possible from the surface before staining has a chance to occur. Allow muddy tracks to dry out, then brush along the grain and vacuum.

Rugs

Like carpets, rugs should be vacuumed regularly to remove dirt and debris which might degrade the pile. Vacuum gently in the direction of the pile, avoiding the fringes. Sweep fringes gently with a brush.

Sources

Hard floors

Bisazza UK
020 8640 7994
www.bisazza.it
Mosaic products using glass and glass agglomerates, mixing traditional Venetian style with new products.

Candy Tile Company
Heathfield
Newton Abbot
Devon TQ12 6RF
01626 831459 for stockists
www.candytiles.co.uk
Victorian, traditional, ethnic and modern ceramic tiles.

Casbah Tiles
20 Wellington Lane
Montpelier
Bristol BS6 5PY
0117 9427318
Handmade cement-based floor tiles from North Africa.

Castelnau Tiles
175 Church Road
London SW13 9HR
020 8741 2452
www.castelnautiles.co.uk
Stone, terracotta and mosaic.

Kenneth Clark Ceramics
The Tile Factory
Unit 4, The Bridges
Business Park
Horsehay
Shropshire TF4 3EE
01952 505085 for catalogue and information
www.kennethclarkceramics.co.uk
Hand-decorated ceramic tiles and borders. Individual commissions undertaken.

Classical Flagstones
Showrooms:
Lower Ledge Farm
Dyrham, Wiltshire SN14 8EY
And
Roadside Court
Alderley Road, Chelford
Cheshire SK11 9AP
0117 9371960 for information and brochure
www.classical-flagstones.com
Stone tiles, cobbles and flags.

Cosmo Ceramics
515 Lawmoor Street
Dixons Blazes Industrial Estate
Glasgow G5 0TY
0141 4201122
Distributors of ceramic, marble, granite and slate tiles.

Craven Dunnill & Co.
Stourbridge Road
Bridgnorth
Shropshire WV15 6AS
01746 761611
www.cravendunnill.co.uk
Large showrooms of ceramic tiles, plus tile adhesives and tiling tools.

CSW Tiling
22–24 Nuffield Road
Nuffield Trading Estate
Fleetsbridge
Poole, Dorset BH17 0RB
01202 675836
Distributor with showrooms of ceramic floor tiles, mosaics.

Delabole Slate
Pengelly, Delabole
Cornwall PL33 9AZ
01840 212242
www.delaboleslate.com
Slate floor tiles and slabs.

Domus Tiles
1 Canterbury Court
6 Camberwell New Road
London SE5 0TG
020 7091 1500
www.domustiles.com
A huge range of tiles in stone, terracotta, ceramic, steel and glass, plus unusual ranges featuring marble pebbles suspended in resin. Design services available.

European Heritage
48–52 Dawes Road
London SW6 7EN
020 7381 6063
www.europeanheritage.co.uk
Natural stone specialist, plus wooden flooring. Showrooms across London.

Feelystone
Kilkenny Limestone Ltd
Kellymount Quarries
Paulstown
Co. Kilkenny
Eire
+353 (0) 5997 26191
www.feelystone.com
Limestone and granite.

Fired Earth
Twyford Mill
Oxford Road, Adderbury
Oxfordshire OX17 3SX
01295 814315 for sales
01295 814300 for brochure
www.firedearth.co.uk
Terracotta and encaustic tiles, including handmade tiles, slate, Venetian marble, limestone and quarry tiles. Also wooden flooring, natural floor coverings and rugs. Showrooms across the UK.

Froyle Tiles
Froyle Pottery
Lower Froyle, Alton
Hampshire GU34 4LL
01420 23693 for information and stockists
www.froyletiles.co.uk
Handmade vitrified stone- and earthenware tiles. Wide range of colours, shapes and sizes.

Gooding Aluminium
1 British Wharf
Landmann Way
London SE14 5RS
020 8692 2255
www.goodingalum.com
Aluminium sheet flooring.

H&L Marble
Units 2–3, Abbey Wharf
Mount Pleasant
Alperton, Middlesex HA0 1NR
0800 7313125
www.hlmarble.co.uk
Granite, limestone and marble.

Hanson Bath and Portland Stone
Bumpers Lane
Portland
Dorset DT5 1HY
01305 820207
Bath, Portland and Cotswold stone.

Hard Rock Flooring
Fleet Marston Farm
Fleet Marston
Aylesbury
Buckinghamshire HP18 0PZ
01296 658755
www.hardrockflooring.co.uk
Terracotta, slate and stone flooring.

Ibstock Building Products
0870 9034000 for nearest sales office
www.ibstock.uk.com
Brick and clay pavers.

Kirkstone
Skelworth Bridge
Ambleside
Cumbria LA22 9NN
01539 433296
www.kirkstone.com
Slate, granite, limestone, terracotta and glass tiles.

Limestone Gallery
Arch 47
South Lambeth Road
London SW8 1SS
020 7735 8555
Antique and modern stone flooring.

Lloyd of Bedwyn
91 Church Street
Great Bedwyn
Marlborough
Wiltshire SN8 3PF
01672 870234
www.lloydofbedwyn.net
Marble, granite and stone for flooring and surfaces.

Mandarin
Unit 1, Wonastow
Industrial Estate
Monmouth NP25 5JB
01600 715444
www.mandarinstone.com
Slate, stone, terracotta and mosaic tiles in many finishes. Further showrooms in Cardiff, Bath and Cheltenham.

Marble Arch
431 & 432 Gordon Business Centre, Gordon Grove
London SE5 9DU
020 7738 7212
www.marblearch.com
Supply, installation and maintenance of marble, granite, limestone and slate.

Mosaic Workshop
1a Princeton Street
London WC1R 4AX
020 7831 0889
www.mosaicworkshop.com
Design and make glass, marble and ceramic mosaics.

Original Style
Falcon Road
Sowton Industrial Estate
Exeter EX2 7LF
01392 473000 for stockists
www.originalstyle.co.uk
Manufacturers of reproduction Victorian geometric and decorative tiles, stone tiles and mosaics, plus ceramics.

The Original Tile Company
23a Howe Street
Edinburgh EH3 6TF
0131 5562013
Terracotta, limestone, slate and marble, plus Victorian and Moroccan encaustic tiles.

Paris Ceramics
583 King's Road
London SW6 2EH
020 7371 7778
www.parisceramics.com
English and French limestone, 18th-century stone and terracotta, hand-cut mosaics and decorative tiles.

Pisani
Unit 12, Transport Avenue
Great West Road
Brentford TW8 9HF
020 8568 5001
www.pisani.co.uk
Granite, marble, limestone and sandstone.

J. Preedy and Sons
Lamb Works, North Road
London N7 9DP
020 7700 0377
www.preedyglass.com
Design and install glass floors.

Reed Harris
Showrooms:
Riverside House
27 Carnwath Road
London SW6 6JE
020 7736 7511
And
Sergeants Industrial Estate
102 Garratt Lane
London SW18 4DJ
020 8877 9774
www.reedharris.co.uk
Ceramic tiles, mosaic, marble and granite.

The Reject Tile Shop
178 Wandsworth Bridge Road
London SW6 2UQ
020 7731 6098
And
Criterion #2 Tiles
2 and 2A Englands Lane
Belsize Park
London NW3 4TG
020 7483 2608
www.criterion-tiles.co.uk
Bargain end-of-line tiles, from terracotta to mosaic.

Dennis Ruabon
Hafod Tileries, Ruabon
Wrexham LL14 6ET
01978 843484
www.dennisruabon.co.uk
Quarry tiles and clay pavers.

Stone Age
Unit 3, Parsons Green Depot
Parsons Green Lane
London SW6 4HH
020 7384 9090
And
14 King's Road
Clifton, Bristol BS8 4AB
0117 9238180
www.stone-age.co.uk
Limestone, sandstone, basalt, granite and marble flooring.

Stonell
08000 832283 for nearest showroom
www.stonell.co.uk
Quality natural stone, marble and mosaic floor tiles.

Sussex Terracotta
Aldershaw Handmade Tiles
Pokehold Wood, Kent Street
Sedlescombe, Battle
East Sussex TN33 0SD
01424 756777
www.aldershaw.co.uk
Handmade clay floor tiles, keystones and borders. Custom-design service.

Terra Firma Tiles
70 Chalk Farm Road
London NW1 8AN
020 7485 7227
And
High Street, Stockbridge
Hampshire S020 6HF
01264 810315
www.terrafirmatiles.co.uk
Natural materials, specialized ceramic tiles and mosaics.

Terranova
Lings Farm, York Road
Bishop Brinton
East Yorkshire HU17 7RU
01964 551555
A wide range of tiles in terracotta, marble, stone, slate, encaustic and ceramic.

Tower Ceramics
91 Parkway
London NW1 7PP
020 7485 7192
www.towerceramics.co.uk
Retailers of all sizes of tiles, glazed and unglazed finishes, terracotta and natural stone.

Anna Wyner
2 Ferry Road
London SW13 9RX
020 8748 3940
Designer of wall and floor mosaics, also terrazzo panels with integral mosaic designs.

Wood floors

Campbell Marson & Co.
573 King's Road
London SW6 2EB
020 7371 5001
www.campbellmarson.com
Solid and laminated hardwood flooring.

Bernard Dru Oak
Bickham Manor
Timberscombe, Minehead
Somerset TA24 7UA
01643 841312
www.oakfloor.co.uk
Specialist in oak flooring.

Ebony and Co.
198 Ebury Street
London SW1W 8UN
020 7259 0000
www.ebonyandco.com
Specialist in handcrafted solid wood floors.

Element 7
Unit 2, Parsons Green Depot
Parsons Green Lane
London SW6 4HH
020 7736 2366
www.element7.co.uk
Luxurious wide-plank floors.

English Timbers
01377 229301
www.englishtimbers.co.uk
Solid timber and hardwood.

Evans Hardwood Flooring
0800 074 5965
www.evanshardwood flooring.co.uk
Supply, fit and finish solid hardwood flooring.

The Hardwood Flooring Company
31–35 Fortune Green Road
London NW6 1DU
020 7341 7000
www.hardwoodflooring company.com
Solid and veneered, unfinished or finished wood flooring.

Junckers
01376 534700 for stockists
www.junckers.co.uk
Danish producer of solid hardwood flooring.

Khars
01243 778747 for stockists
www.kahrs.com
Fine Swedish wood flooring.

Lassco Flooring
Maltby Street
London SE1 3PA
020 7394 2101
www.lassco.co.uk/flooring
New and reclaimed hardwood, including oak, wood block, parquet and Victorian pine boards. Also flagstones and terracotta.

Orlestone Oak Saw Mill
Nickley Wood, Shadoxhurst
Ashford, Kent TN26 1LZ
01233 732179
www.orlestoneoak.co.uk
Specialist in oak flooring, made to specification.

Parquet and General Flooring Co.
Grange Lane, Winsford
Cheshire CW7 2PS
01606 861442
www.wideboards.com
Design and installation of bespoke hardwood floors.

Panda Flooring Company
1 Grange Park, Thurnby
Leicester LE7 9QQ
0116 2414816
www.pandaflooring.co.uk
Bamboo hardwood flooring.

Priors Reclamation
Unit 65, Ditton Priors Trading Estate
Ditton Priors, Bridgenorth
Shropshire WV16 6SS
01746 712450
www.priorsrec.co.uk
New and reclaimed solid timber flooring.

Solid Floor
53 Pembridge Road
London W11 3HG
020 7221 9166
www.solidfloor.co.uk
Solid timber floors. Showrooms across London, in Edinburgh and Glasgow.

Tarkett
01905 342743 for nearest dealer
www.tarkett-floors.com/uk
Wood, laminate and vinyl flooring, including Marley Floors (see separate entry).

Upofloor UK
Brook Farm
Horsham Road, Cowfold
West Sussex RH13 8AH
01403 860000
www.upofloor.co.uk
Wooden floor systems from Finland.

Victorian Woodworks
54 River Road, Creek Mouth
Barking, Essex IG11 0DW
020 8534 1000
www.victorianwoodworks.co.uk
Manufacturers and suppliers of solid new hardwood, reclaimed wood and antique timber flooring.

Victorian Woodworks Contracts
The Flooring Studio
158 Walton Street
London SW3 2JL
020 7225 3240
www.victorianwoodworkcontracts.co.uk
Supply, lay and treat hard, wooden and soft floorings.

Walcot Reclamation
108 Walcot Street
Bath BA1 5BG
01225 444404
www.walcot.com
Salvaged and re-sawn oak flooring, woodblock and quarry tiles.

Watco
01483 418418 for information and sales
www.watco.co.uk
Resins and oils for wood floors, and other flooring treatments.

Weldon Flooring
The Grange, Norton Disney
Lincolnshire LN6 9JP
01636 894838
www.weldon.co.uk
Design, supply and install hardwood flooring, specializing in the restoration of marquetry and parquet.

The West Sussex Antique Timber Company
Reliance Works, Newpound
Wisborough Green
West Sussex RH14 0AZ
01403 700139
www.wsatimber.co.uk
Specialists in supplying and laying antiqued oak flooring.

Sheet & soft tile

Alma Home
8 Vigo Street
London W1S 3HJ
020 7439 0925 for showroom
020 7377 0762 for information
www.almahome.co.uk
Specialists in skins for rugs, flooring and furnishings. Also available from Selfridges (see separate entry).

Altro
01462 480480 for stockists
www.altro.co.uk
Vinyl, non-slip PVC and rubber sheet and tile flooring.

Bill Amberg
10 Chepstow Road
London W2 5BD
020 7727 3560
www.billamberg.com
Leather flooring.

Amtico
0800 667766 for stockists and showrooms
www.amtico.com
Durable flooring that resembles natural materials, plus glass, metallics and other special finishes. Floors can be designed and laid to your specifications.

Benson Design
121 Walm Lane
London NW2 4QE
020 8452 8864
www.bensondesign.co.uk
Handmade leather rugs and floors.

Bonar Floors
01773 744121 for information and brochure
www.bonarfloors.com
www.flotex.co.uk
Suppliers of Flotex high-specification vinyl flooring.

Bragman Flett
020 8337 1934 for details
www.bragmanflett.co.uk
Sheet metal flooring.

Dalsouple
01278 727733 for stockists
www.dalsouple.com
Manufacturers of rubber flooring in a huge range of colours and textures, including marbled and terazzo finishes, plus rubber flooring for outdoors.

First Floor
174 Wandsworth Bridge Road
London SW6 2UQ
020 7736 1123
www.firstfloor.uk.com
A contemporary range of lino, vinyl, rubber and cork flooring, plus carpets. Suppliers of Dalsouple (see separate entry).

Forbo-Nairn
01592 643111 for stockists
www.forbo-flooring.co.uk
Leading manufacturer of lino and cushioned vinyl flooring, including Novilon, Novilux and Marmoleum (see separate entry).

Freudenberg Building Systems
Unit 6, Wycliffe Industrial Park
Leicester Road, Lutterworth
Leicestershire LE17 4HG
01455 204483
www.nora.com
Rubber tiles and sheet flooring.

Gerflor
01926 401500 for stockists
www.gerflor.com
Vinyl, PVC and specialist sheet and tile flooring.

Harvey Maria
020 8542 0088 for sales and brochure
www.harveymaria.co.uk
Vinyl tiles with contemporary photographic images and more traditional textured stone finishes.

Jaymart Rubber & Plastics
01373 864926 for information and sales
www.jaymart.co.uk
Vinyl and rubber flooring.

Karndean International
Showroom:
Crabapple Way
Vale Park, Evesham
Worcestershire WR11 1GP
01386 820100 for information and stockists
www.karndean.co.uk
Heavy-duty customized vinyl flooring in tile and plank form.

Marley Floors
01622 85040 for information and brochures
www.marleyfloors.com
Wide range of vinyl flooring.

Marmoleum
01592 643777
www.marmoleum.co.uk
Linoleum in different colours, patterns and textures.

Polyflor
0161 7671111 for information and distributors
www.polyflor.com
Sheet and tile vinyl and rubber specialists.

Siesta Cork Tile Company
020 8683 4055 for stockists and sales
www.siestacorktiles.co.uk
Cork tiles in a variety of colours and finishes. From stockists or by mail order.

Sinclair Till Flooring
791–793 Wandsworth Road
London SW8 3JQ
020 7720 0031
www.sinclairtill.co.uk
Lino, natural matting, rubber, cork, vinyl and carpet.

Wicanders
01403 710001 for stockists
www.wicanders.com
Easy-maintenance, good-looking cork floors in many colours and patterns. Cork tiles with wood veneers and cork/rubber composite tiles.

Soft floors

Allied Carpets
08000 932932 for showrooms
www.alliedcarpets.com
A wide range of carpets, plus rugs, vinyls, laminates and wood floors.

The Alternative Flooring Company
01264 335111 for stockists
www.alternative-flooring.co.uk
Natural-fibre floor coverings, including sisal, coir, seagrass, jute and wools.

Anglo Felt Industries
Tong Lane
Whitworth
Rochdale
Lancashire OL12 8BG
01706 853513 for information and sales
www.anglofelt.com
Carpet underfelt in wool, hair/jute and felt/rubber blends, and double-stick varieties.

Axminster Carpets
Woodmead Road
Axminster
Devon EX13 5PQ
01297 630650 for stockists
www.axminstercarpets.co.uk
Makers of Axminster, Wilton and tufted carpets with an individual manufacturing and design service.

David Black Carpets
27 Chepstow Corner
Chepstow Place
London W2 4XE
020 7727 2566
www.david-black.com
Turkish hand-knotted carpets made to order in wool pile or kilim flatweave, plus a range of vegetable-dyed cotton dhurries.

Bosanquet Ives
World's End Studios
132–134 Lots Road
London SW10 0RJ
020 7349 7042
www.bosanquet-ives.co.uk
Bespoke and off-the-shelf carpets, rugs and runners.

Brintons
0800 505055 for brochure and stockists
www.brintons.net
Manufacturers of quality woven carpets, including Axminsters, Wiltons and tufted ranges.

The Conran Shop
Michelin House
81 Fulham Road
London SW3 6RB
020 7589 7401
And
55 Marylebone High Street
London W1U 5HS
020 7723 2223
www.conran.co.uk
Contemporary rugs and mats.

Craigie Stockwell Carpets
81 York Street
London W1H 1QH
020 7224 8380
www.craigiestockwell carpets.com
Custom-design service for carpets, plus motifs and borders in modern or classic designs. Can also match antique carpets.

Crucial Trading
01562 743747 for stockists and showrooms
www.crucial-trading.com
Over 180 textures, colours and styles of floor covering in sisal, paper, bamboo, jute, seagrass, coir and wool, available in runners, matting and wall-to-wall.

Curragh Tintawn Carpets
01297 630647 for retailers
Traditional Irish country-house-style carpets in textured combinations of twist and velvet pile.

Custom Carpet Company
01737 830301
www.customcarpet company.co.uk
Individually designed carpets, from concept to fitting.

Christopher Farr Handmade Rugs
6 Bunsall Street
London SW3 3ST
020 7349 0888
www.cfarr.co.uk
Specialist in contemporary designs, including rugs by Romeo Gigli, Rifat Ozbek and Allegra Hicks.

Fineweave Carpets
Unit 18, Headley Park Ten
Headley Road East
Woodley, Berkshire RG5 4SW
01494 534620
Wide range of British and European carpets.

Fired Earth
See listing under Hard floors
Rugs and natural fibres including coir, jute and sisal.

The Flokati Rug Company
Unit 12, The Osiers Estate
Enterprise Way
London SW18 1EJ
020 8337 3005
www.flokatirugco.co.uk
Pure wool flokati rugs and wool-mix Greek kilims.

Gideon Hatch Rugs and Carpets
1 Port House
Plantation Wharf
London SW11 3TY
020 7223 3996
www.gideonhatch.co.uk
Handspun woollen rugs in natural colours.

Gilt Edge Carpets Ltd
255 New King's Road
London SW6 4RB
020 7731 2588
www.giltedgecarpets.co.uk
Specialist in wool and natural carpets, plus vinyl and wood floors. Planning/fitting service.

Habitat
196 Tottenham Court Road
London W1T 7LG
020 7631 3880 for shop
0870 4115501 for customer services
www.habitat.net
Contemporary rugs. Branches nationwide.

Hart of the House
The Old White Hart
Long Melford
Sudbury
Suffolk CO10 9HX
www.hartofthehouse.co.uk
Contemporary rugs and runners in natural materials.

Heal's
196 Tottenham Court Road
London W1P 9LD
020 7636 1666
www.heals.co.uk
Classic contemporary rugs. Branches in Chelsea, Manchester, Kingston and Guildford.

Interior Supply
2/16 Chelsea Harbour
Design Centre
London SW10 0XE
020 7352 0502
www.interiorsupply.co.uk
Natural floor coverings and rugs.

Katja Designs
240 Brompton Road
London SW3 2BB
020 7584 9914
Rugs in wool, cotton and paper yarn.

Kersaint Cobb & Co.
01675 430430 for stockists
www.kersaintcobb.co.uk
Natural and pure wool floor coverings.

Kilim-warehouse.com
(+34) 616 512 209 for brochure
www.kilim-warehouse.com
Wide selection of kilims and flatweave rugs from e-shop based in Spain.

Liberty
214 Regent Street
London W1B 5AH
020 7734 1234
www.liberty.co.uk
Oriental rugs.

Loomah
Redloh House
2 Michael Road
London SW6 2AD
020 7371 9955
www.loomah.com
Design bespoke handmade carpets and rugs.

Hugh Mackay
0191 3784444
www.hughmackay.co.uk
Durham-based manufacturer of Axminster, Wilton and tufted carpets, plus bespoke service in the quality ranges.

Newhey Carpets
Design studio and showroom:
Gordon Street
Newhey
Rochdale
Lancashire OL16 3SL
01706 846375 for sales
Axminster and tufted carpets. Bespoke service.

Annette Nix
150a Camden Street
London NW1 9PA
020 7482 7041
Contemporary rug designer.

Roger Oates Design
1 Munro Terrace
London SW10 0DL
020 7351 2288
And
The Long Barn
Eastnor
Herefordshire HR8 1EL
01531 631611 for stockists
www.rogeroates.com
A timeless collection of flatweave runners, rugs and fabrics.

Penthouse Carpets
Buckley Carpet Mill
Buckley Road
Rochdale
Lancashire OL12 9DU
01706 341231
Specialist manufacturer of twist-pile carpets.

The Rug Studio
34 High Street
Hampton Wick
Kingston upon Thames
KT1 4DB
020 8977 4403
www.therugstudio.co.uk
Antique and traditional rugs from around the world.

Ryalux
01706 716000
www.ryalux.com
Ryalux and Lomas carpets, plus ranges inspired by the V&A and Chatsworth House.

Selfridges
400 Oxford Street
London W1A 1AB
08708 370470 for customer services
www.selfridges.co.uk
Rugs and carpets.

Stairrods
01207 591543 for stockists
www.stairrods.co.uk
Rods in traditional and contemporary designs. Bespoke service available.

Anthony Thompson
7 Kensington Park Gardens
London W11 3HB
www.anthonythompsonltd.co.uk
Traditional and contemporary rugs, plus bespoke service.

Top Floor
2/6 Chelsea Harbour
Design Centre
Chelsea Harbour
London SW10 0XE
020 7795 3333
www.topfloorrugs.com
Contemporary rugs and carpets.

Ulster Carpets Mills
Castleisland Factory
Craigavon BT62 1EE
Northern Ireland
028 3833 4433
www.ulstercarpets.co.uk
Axminster, Wilton and tufted carpets in extensive range of designs and qualities, including Cath Kidston range.

Christine Vanderhurd
By appointment:
Studio 276
56 Gloucester Road
London SW7 4UB
020 7313 5400
Contemporary rug designer.

Veedon Fleece
By appointment:
42 Nightingale Road
Guildford, Surrey GU1 1EP
01483 575758
www.veedonfleece.com
Hand-knotted carpets in traditional and contemporary designs.

Victoria Carpets
Worcester Road
Kidderminster
Worcestershire DY10 1HL
01562 749349
Manufacturers of woven and tufted carpets in wool-rich yarns.

Vigo Carpet Gallery
6a Vigo Street
London W1S 3HF
020 7439 6971
Specialize in large carpets, new needlework and tufted rugs.

Waveney Rush Industry
The Old Maltings
Caldecott Road
Oulton Broad, Lowestoft
Suffolk NR32 3PH
01502 538777
www.waveneyrush.co.uk
Handmade rush-weave floor coverings.

The Wilding Partnership
The Malt House, Eardisley
Herefordshire HR3 6NH
01544 327405
www.wildingpartnership.co.uk
Designers and manufacturers of hand-tufted rugs and carpets, also specialize in carpets and rugs produced from original artwork.

Woodward Grosvenor
Stourvale Mills
Green Street, Kidderminster
Worcestershire DY10 1AT
01562 820020
Manufacturers of high-quality Axminster and Wilton carpet, bespoke service for archive designs dating back to 1790.

Helen Yardley
28–29 Great Sutton Street
London EC1V 0DS
020 7253 9242
www.helenyardley.com
Leading designer of contemporary rugs.

General information

British Interior Design Association
3/18 Chelsea Harbour
Design Centre
London SW10 0XE
020 7349 0800
www.bida.org
Advice in locating a professional interior designer.

British Wood Preserving and Dampproofing Association
01332 225100
www.bwpda.co.uk
Help you find a contractor and offer information on all aspects of wood treatment and damp control.

British Wool Marketing Board
01274 688666
www.britishwool.org.uk
Technical information, advice and publications.

The Building Centre
26 Store Street
London WC1E 7BT
020 7692 4000
09065 161136 for charged information line
Comprehensive, independent information, and permanent exhibits of materials and products from a vast range of suppliers.

Business Design Centre
52 Upper Street
London N1 0QH
020 7359 3535
www.businessdesigncentre.co.uk
Over 100 trade and contract showrooms; also hosts regular exhibitions.

The Carpet Foundation
0845 6012200
www.carpetfoundation.com
Technical information, advice and publications.

Carpet Information Centre
www.carpetinfo.co.uk
Advice about carpets.

Crafts Council
44a Pentonville Road
London N1 9BY
020 7278 7700
www.craftscouncil.org.uk
Promotes contemporary crafts; regular exhibitions.

Federation of Master Builders
Gordon Fisher House
14–15 Great James Street
London WC1N 3DP
020 7242 7583
www.fmb.org.uk
16,000 members in the construction industry, 11 regional offices, and 140 local branches.

Forest Stewardship Council
Unit D, Station Buildings
Llanidloes
Powys SY18 6EB
01686 413916
www.fsc-uk.info
International body certificating timber from sustainable sources around the world.

National Home Improvement Council
020 7828 8230
www.nhic.org.uk
Advice and information.

National Institute of Carpet and Floorlayers
0115 9583077
www.nicfltd.org.uk
Listings of qualified fitters in your area.

Royal Institute of British Architects
66 Portland Place
London W1B 1AD
020 7580 5533
www.riba.org
For general advice and information on architectural practice.

The Tile Association
Forum Court
83 Copers Cope Road
Beckenham
Kent BR3 1NR
020 8663 0946
www.tiles.org.uk
Represent all aspects of the UK wall- and floor-tile industry, offering information and listings of members.

General suppliers

IKEA
0845 3551141 for nearest branch
www.ikea.com
Contemporary rugs, mats, runners, laminate and wood floors, tiles.

John Lewis Partnership
08456 049049 for nearest branch
www.johnlewis.com
Comprehensive range of carpet, rugs, and sheet and tile flooring.

Index

Picture credits

Architects and designers whose work is featured in this book

23 Architecture
S.I. Robertson
318 Kensal Road
London W10 5BZ
020 8962 8666
fax 020 8962 8777
stuart@23arc.com
www.23arc.com
Pages 34–35

Adjaye Associates
(formerly Adjaye & Russell)
23–28 Penn Street
London N1 5DL
020 7739 4969
fax 020 7739 3484
www.adjaye.com
Pages 8a, 26–27

Agnes Emery
Emery & Cie
Noir D'Ivoire
Rue de l'Hôpital 25–29
1000 Bruxelles
+32 2 513 58 92
fax +32 2 513 39 70
Moroccan tiles, concrete floor tiles, selected paints.
Pages 22ar, 59ar

Alidad Ltd
2 The Lighthouse Gasworks
Michael Road
London SW6 2AD
020 7384 0121
fax 020 7384 0122
www.alidad.com
Page 151r

Amanda Freedman
based in Notting Hill,
London
fax 020 7727 6860
Pages 90, 91ac, 91bl

Andrew Parr
SJB Interior Design Pty Ltd
25 Coventry Street
South Melbourne
Australia
+61 3 9686 2122
Pages 13bl, 78ar, 80a, 139br

Angela A'Court
Artist
orangedawe@hotmail.com
Pages 34–35

Ann Boyd Design Ltd
33 Elystan Street
London SW3 3NT
020 7591 0202
Pages 133, 144–145

Ash Sakula Architects
24 Rosebery Avenue
London EC1R 4SX
020 7837 9735
fax 020 7837 9708
info@ashsak.com
www.ashsak.com
Pages 28, 44al, 130l, 156–7

behun/ziff design
153 East 53rd Street,
43rd Floor
New York, NY 10022
+1 212 292 6233
fax +1 212 292 6790
Page 15ar

Belmont Freeman Architects
110 West 40 Street
New York, NY 10018
+1 212 382 3311
fax +1 212 730 1229
www.belmontfreeman.com
Page 105l

Ben Kelly Design
10 Stoney Street
London SE1 9AD
020 7378 8116
fax 020 7378 8366
www.benkellydesign.com
Page 83b

buildburo ltd
7 Tetcott Road
London SW10 0SA
020 7352 1092
fax 020 7351 3986
www.buildburo.co.uk
Page 141bc

Campion A. Platt
152 Madison Avenue,
Suite 900
New York, NY 10016-5424
+1 212 779 3835
fax +1 212 779 3851
www.campionplatt.com
Page 49

Carla Saibene
Shop:
Carla Saibene
via San Maurilio 20
Milano
Italy
tel/fax +39 2 77 33 15 70
xaibsrl@yahoo.com
Womenswear collection, accessories and antiques.
Page 60l

Carnachan Architects Ltd
33 Bath Street
P.O. Box 37-717
Parnell, Auckland
New Zealand
+64 9 3797 234
fax +64 9 3797 235
Page 12l

Carole Oulhen
Interior Designer
+33 6 80 99 66 16
fax +33 4 90 02 01 91
Pages 51a, 51b

Charles Bateson Design Consultants
18 Kings Road, St Margaret's
Twickenham TW1 2QS
020 8892 3141
fax 020 8891 6483
Charles.bateson@btinternet.com
Page 137a

Charles Rutherfoord
51 The Chase
London SW4 0NP
020 7627 0182
Pages 21ar, 30bl, 31r, 50br, 68–69, 69a, 75bl, 75r, 76b, 83al, 83ac, 84, 86, 123 both, 128–9, 147bl

Charlotte Barnes Interiors
26 Stanhope Gardens
London SW7 5QX
020 7244 9610
Page 93

Christophe Gollut
Alistair Colvin Limited
116 Fulham Road
London SW3 6HU
020 7370 325
www.christophegollut.com
Pages 50–51a, 102l

Circus Architects
Unit 111 The Foundry
165 Blackfriars Road
London SE1 8EN
020 7953 7322
fax 020 7953 7255
Pages 14–15, 29r, 30br, 119

Colin Orchard Design
219a King's Road
London SW3 5EJ
07002 753762
Page 143

DAD Associates
112–16 Old Street
London EC1V 9BD
020 7336 6488
Pages 11ar, 11br, 21bl, 62–63, 63l, 83ar, 91ar

Daniel Jasiak
Designer
12 rue Jean Ferrandi
Paris 75006
+33 1 45 49 13 56
fax +33 1 45 49 23 66
Pages 88al, 88ar

David Khouri
Comma
149 Wooster Street,
Suite 4NW
New York, NY 10012
+1 212 420 7866
fax +1 212 202 3584
info@comma-nyc.com
www.comma-nyc.com
Architecture, interiors and furniture.
Page 71

Emulsion
172 Foundling Court
Brunswick Centre
London WC1N 1QE
020 7833 4533
fax 020 7278 9975
contact@emulsion architecture.com
www.emulsion architecture.com
Page 96

Eric De Queker
DQ — Design In Motion
Koninklijkelaan 44
2600 Bercham
Belgium
Pages 11ac, 18b

Featherstone Associates
74 Clerkenwell Road
London EC1M 5QA
020 7490 1212
fax 020 7490 1313
sarah.f@featherstone-associates.co.uk
www.featherstone-associates.co.uk
Pages 25ar, 43, 111, 136–137

Felix Bonnier
7 rue St Claude
75003 Paris
+33 1 42 26 09 83
Pages 22b, 27a, 67, 79, 103, 145a

Filer & Cox
40 Stockwell Street
Greenwich
London SE10 8EY
020 8269 2211
fax 020 8269 2205
www.filerandcox.com
Pages 105r, 107, 118l

François Muracciole
Architect
54, rue de Montreuil
75011 Paris
+33 1 43 71 33 03
francois.muracciole@libertysurf.fr
Page 59al

Frédéric Méchiche
4 rue de Thorigny
75003 Paris
+33 1 42 78 78 28
Pages 4–5, 16–17, 29al, 40l, 41, 52–3, 53l, 74–5, 89, 138–139, 141a

Gabellini Associates
665 Broadway, Suite 706
New York, NY 10012
+1 212 388 1700
fax +1 212 388 1808
www.gabelliniassociates.com
Pages 32–33 main

Gavin Jackson Architects
07050 097561
Page 73l

Guard Tillman Pollock
(Mark Guard Architects)
161 Whitfield Street
London W1T 5ET
020 7380 1199
fax 020 7387 5441
www.markguard.com
Pages 18c, 29bl, 68r, 91br

Helen Ellery of
The Plot London
77 Compton Street
London EC1V 0BN
020 7251 8116
fax 020 7251 8117
helen@theplotlondon.com
www.theplotlondon.com
Pages 22ac, 45ar, 58ac, 101

Helm Architects
2 Montagu Row
London W1U 6DX
020 7224 1884
fax 020 7224 1885
nh@helmarchitects.com
Page 22ar

Hudson Architects
49–59 Old Street
London EC1V 9HX
020 7490 3411
fax 020 7 490 3412
anthonyh@hudson architects.co.uk
www.hudsonarchitects.co.uk
Pages 18al, 21cr, 25bl, 42–43, 44ar, 46–47, 80c, 80–81

Hugh Broughton Architects
4 Addison Bridge Place
London W14 8XP
020 7602 8840
fax 020 7602 5254
www.hbarchitects.co.uk
Pages 104–105, 112

Ilaria Miani
Shop:
via Monserrato 35
00186 Roma
+39 6 68 33 160
ilariamiani@tin.it
Podere Buon Riposo in Val d'Orcia is available to rent.
Page 36r

Imogen Chappel
07803 156081
Page 12ar

Interior Concepts
6 Warren Hall
Manor Road, Loughton
Essex IG10 4RP
020 8508 9952
fax 020 8502 4382
www.jointeriorconcepts.co.uk
Pages 132r, 134l

Interni Pty Ltd.
98 Barcom Avenue
Rushcutters Bay
NSW 2011
Australia
+61 2 9360 5660
fax +61 2 9331 3104
Pages 22al, 56, 56–57a, 57b, 78bl, 78br, 160

IPL Interiors
François Gilles and Dominique Lubar
25 Bullen Street
London SW11 3ER
020 7978 4224
fax 020 7978 4334
Pages 40c, 44ac, 45al, 48l, 95, 145b, 152r, 155br

Jacqueline Coumans
Le Décor Français
1006 Lexington Avenue
New York, NY 10021
+1 212 734 0032
fax +1 212 988 0816
www.ledecorfrancais.com
info@ledecorfrancais.com
Page 92bl

James Biber, AIA
Pentagram Architecture
204 Fifth Avenue
New York, NY 10010
www.pentagram.com
Page 78al

Jane Churchill Interiors
81 Pimlico Road
London SW1W 8PH
020 7730 8564
fax 020 7823 6421
janechurchill@jcildircon.co.uk
Page 100r

Jennifer Post Design
390 West End Avenue
New York, NY 10024
+1 212 769 0338
Page 74l

JoAnn Barwick Interiors
P.O. Box 982
Boca Grande
Florida 33921
Pages 99bl, 99ar

Johanne Riss
35 Place du Nouveau Marché aux Graens
1000 Bruxelles
+32 2 513 0900
fax +32 2 514 3284
www.johanneriss.com
Stylist, designer and fashion designer.
Page 64ar

John Barman Inc.
500 Park Avenue
New York, NY 10022
+1 212 838 9443
john@barman.com
www.johnbarman.com
Interior design and decoration.
Pages 115l, 116l, 159

John C. Hope Architects
3 St Bernard's Crescent
Edinburgh EH4 1NR
0131 315 2215
fax 0131 315 2911
Page 88b

John F. Saladino
Saladino Group Inc
200 Lexington Avenue,
Suite 1600
32nd–33rd Street
New York, NY 10016
www.saladinofurniture.com
+1 212 684 6805
fax +1 212 684 3753
Page 85a

John Minshaw Designs Ltd
17 Upper Wimpole Street
London W1H 6LU
020 7258 5777
fax 020 7486 6777
enquiries@johnminshawdesigns.com
Pages 77a both, 86–87

Johnson Naylor
13 Britton Street
London EC1M 5SX
020 7490 8885
fax 020 7490 0038
www.johnsonnaylor.com
Page 148r

Just Design Ltd
80 Fifth Avenue, 18th Floor
New York, NY 10011
+1 212 243 6544
fax +1 212 229 1113
wbp@angel.net
Page 92a

Karim Rashid Inc.
357 West 17th Street
New York, NY 10011
+1 212 929 8657
fax +1 212 929 0247
www.karimrashid.com
Industrial design.
Page 70l

Khai Liew Design
166 Magill Road
Norwood
South Australia 5067
+61 8 8362 1076
Pages 37, 55

L.B.D.A. (Laura Bohn Design Associates, Inc.)
30 West 26th Street
New York, NY 10010
+1 212 645 3636
fax +1 212 645 3639
www.lbda.com
Pages 8–9

Lena Proudlock
www.lenaproudlock.com
Page 40r

Luigi Rosselli
Surry Hills, NSW 2010
Sydney, Australia
+61 2 9281 1498
Pages 13a, 21br, 29cl, 36l

Malin Iovino Design
tel/fax 020 7252 3542
iovino@btconnect.com
Page 70br

Marino & Giolito
161 West 16th Street
New York, NY 10011
tel/fax +1 212 675 5737
marino.giolito@rcn.com
Architecture/interior design.
Pages 58c, 58ar, 116r

Marc Meiré
Meirefamily@aol.com
Pages 96–97

Mary Drysdale, Drysdale, Inc.
78 Kalorama Circle NW
Washington, D.C. 20008
+1 202 588 0700
Page 100l

McDowell & Benedetti
68 Rosebery Avenue
London EC1R 4RR
020 7278 8810
fax 020 7278 8844
www.mcdowellbenedetti.com
Pages 13br, 61, 69b, 93, 130br, 130–131, 153

McLean Quinlan Architects
1 Milliners, Riverside Quarter
Point Pleasant
London SW18 1LP
020 8870 8600
fax 020 8870 1567
www.mcleanquinlan.com
Architecture and design.
Pages 25br, 132l

Michael Neumann Architecture
11 East 88th Street
New York, NY 10128
+1 212 828 0407
www.mnarch.com
Page 77br

Mireille and Jean Claude Lothon
La Cour Beaudeval Antiquities
4 rue des Fontaines
28210 Faverolles, France
+33 2 37 51 47 67
Pages 30al, 53r

Moneo Brock Studio
Francisco de Asis
Mendez Casariego 7, Bajo
28002 Madrid
+34 91 563 8056
fax +34 91 563 8573
contact@moneobrock.com
www.moneobrock.com
Page 152l

Mullman Seidman Architects
443 Greenwich Street, # 2A
New York, NY 10013
+1 212 431 0770
fax +1 212 431 8428
www.mullmanseidman.com
Page 92–93

Nelly Guyot, Décoratrice
+33 6 09 25 20 68
Page 54r

OKA Direct
www.okadirect.com
0870 1606002
Page 150bl

Olivier Vidal
14 rue Moncey
75009 Paris
+33 1 49 70 82 82
Pages 8br, 20, 64–65

Paul Collier, Architect
209 rue St Maur
75010 Paris
+33 1 53 72 49 32
paul.collier@architecte.net
Page 75c

Paul Simmons
Timorous Beasties
7 Craigend Place, Anniesland
Glasgow G13 2UN
0141 9593331
Bespoke, hand-printed textiles.
Page 22ar

Philippe Model
33 Place du Marché St. Honoré
75001 Paris
+33 1 42 96 89 02
Decoration, home furnishing and coverings.
Pages 15br, 59bl

Plain English
Stowupland Hall
Stowupland, Stowmarket
Suffolk IP14 4BE
01449 774028
www.plainenglishdesign.com
Cupboard makers.
Page 38 both

Reed Creative Services Ltd
151a Sydney Street
London SW3 6NT
020 7565 0066
fax 020 7565 0067
Pages 76ar, 136l

Retrouvius Reclamation & Design
2A Ravensworth Road
London NW10 5NR
tel/fax 020 8960 6060
mail@retrouvius.com
www.retrouvius.com
Page 22ar

Robert Dye Associates
Linton House
39–51 Highgate Road
London NW5 1RS
020 7267 9388
fax 020 7267 9345
info@robertdye.com
www.robertdye.com
Design consultants/ chartered architects.
Pages 6–7, 18–19, 48c, 74br, 106, 110–111

Roger Oates Design
The Long Barn
Eastnor, Ledbury
Herefordshire HR8 1EL
01531 631611
www.rogeroates.co.uk
Pages 7cl, 24, 25ac, 31l, 54l, 125l, 125r, 127br, 139ar, 140, 144a, 146l, 147r, 154l, 154–155, 155a, 155bl

SCDA Architects
10 Teck Lim Rd
Singapore 088386
+65 6324 5458
fax +65 6324 5450
www.scdaarchitects.com
Page 45ac

Schack-Arnott
Danish Classic Moderne
Andrew Arnott and Karin Schack
517 High Street
Prahan, VIC 3181
Australia
+61 3 9525 0250
Page 149

Sixty 6
66 Marylebone High Street
London W1M 3AH
Pages 158b, 158–159

Stephen Slan AIA
Variations in Architecture
5537 Hollywood Boulevard
Los Angeles, CA 90028
+1 323 962 9101
fax +1 323 962 9127
www.viarc.com
Pages 108–109

Stephen Varady Architecture
P.O. Box 105
St Peters
NSW 2044
Sydney, Australia
+61 2 9516 4044
fax +61 2 9516 4541
Pages 27b, 56–57b, 57ar, 102r

Susan Cropper
www.63hlg.com
Page 105cl

The Amtico Company
Solar Park
Southside
Solihull
West Midlands B90 4SH
0800 667766
www.amtico.com
Pages 120, 121

The Moderns
900 Broadway, Suite 903
New York, NY 10003
+1 212 387 8852
fax +1 212 387 8824
www.themoderns.com
Page 158ar

Tito Canella
Canella & Achilli Architects
via Revere # 7/9
20123 Milano
Italy
+39 2 46 95 222
fax +39 2 48 13 704
ac@planet.it
www.canella-achilli.com
Page 18ar

Todhunter Earle Interiors
Chelsea Reach, 1st Floor
79–89 Lots Road
London SW10 0RN
020 7349 9999
fax 020 7349 0410
interiors@todhunterearle.com
www.todhunterearle.com
Pages 122, 124–125, 128c

Urban Research Lab
www.smcurbanlab.com
Page 118r

Vicente Wolf Associates, Inc.
333 West 39th Street,
Suite 1001
New York, NY 10018
+1 212 465 0590
fax +1 212 465 0639
Pages 63r, 66, 102c, 129

Voon Wong & Benson Saw
Unit 27, 1 Stannary Street
London SE11 4AD
020 7587 0116
fax 020 7840 0178
www.voon-benson.co.uk
Pages 10–11

William Yeoward
270 King's Road
London SW3 5AW
020 7349 7828
www.williamyeoward.com
Page 143

Woolf Architects
4th Floor Studios
39–51 Highgate Road
London NW5 1RT
020 7428 9500
Pages 17l, 25al, 68l, 72–3

Yves Halard
252 Bis Boulevard St. Germain
75007 Paris
+33 1 42 22 60 50
Interior decoration.
Page 127l

Zina Glazebrook
ZG Design
Bridgehampton, New York
+1 631 537 4454
fax +1 631 537 4453
zina@zgdesign.com
www.zgdesign.com
Pages 64br, 141bl

Photography

All photographs by Henry Bourne unless otherwise stated.
Key: ph= photographer, a=above, b=below, r=right, l=left, c=centre.

2 ph Chris Everard/Nadav Kander and Nicole Verity's house; ***3*** ph Tom Leighton; ***4–5*** Frédéric Méchiche's apartment in Paris; ***6–7*** a loft in London designed by Robert Dye Associates; ***7l*** rug by Garouste Bonnetti; ***7cl*** Roger Oates Design; ***7r*** ph James Merrell; ***8al*** ph Tom Leighton; ***8bl*** an apartment in London designed by Adjaye & Russell; ***8br*** an apartment in Paris designed by Olivier Vidal AIA; ***8–9*** ph David Montgomery/Laura Bohn's apartment in New York designed by Laura Bohn Design Associates; ***9ar*** ph Christopher Drake; ***9br*** ph Chris Everard; ***10–11*** ph Christopher Drake/Florence Lim's house in London—architecture by Voon Wong Architects, interior design by Florence Lim Design; ***11ac*** ph Chris Everard/Eric De Queker's apartment in Antwerp; ***11ar&br*** DAD Associates; ***12l*** ph James Merrell/Robyn and Simon Carnachan's house in Adelaide; ***12ar*** ph Debi Treloar/Imogen Chappel's home in Suffolk; ***12br*** Alan and Hepzibah's home in Sussex; ***13a*** ph James Merrell/a house in Sydney designed by Luigi Rosselli; ***13bl*** ph James Merrell/Andrew Parr's house in Melbourne; ***13br*** a mews house in London designed by architects McDowell & Benedetti; ***14*** Alan and Hepzibah's home in Sussex; ***14–15*** Circus Architects; ***15c*** ph Debi Treloar; ***15ar*** ph Chris Everard/the Sugarman-Behun house on Long Island; ***15br*** ph Chris Everard/Philippe Model's apartment in Paris; ***16–17*** Frédéric Méchiche's apartment in Paris; ***17l*** a house in London designed by Woolf Architects; ***18al*** a house in Devon designed by Anthony Hudson Architects; ***18ar*** ph Chris Everard/an apartment in Milan designed by Tito Canella of Canella & Achilli Architects; ***18c*** an apartment in Paris designed by Mark Guard Architects; ***18b*** ph Chris Everard/Eric De Queker's apartment in Antwerp; ***18–19*** a loft in London designed by Robert Dye Associates; ***20*** an apartment in Paris designed by Olivier Vidal AIA; ***21al*** rug by Garouste Bonnetti; ***21bl*** DAD Associates; ***21ar*** Charles Rutherfoord's house in London; ***21cr*** a house in Devon designed by Anthony Hudson Architects; ***21br*** ph James Merrell/a house in Sydney designed by Luigi Rosselli; ***22al*** ph James Merrell/Interni Interior Design Consultancy; ***22ac*** ph Christopher Drake/a house in Salisbury designed by Helen Ellery of The Plot London; ***22ar*** ph Jan Baldwin/a family home in Parsons Green, London: architecture by Nicholas Helm and Yasuyuki Fukuda (architectural assistant) of Helm Architects, interior design and all material finishes supplied by Maria Speake of Retrouvius Reclamation & Design; ***22b*** Felix Bonnier's apartment in Paris; ***23*** ph Andrew Wood/Heidi Kingstone's apartment in London; ***24*** ph Christopher Drake/Fay and Roger Oates' house in Ledbury; ***25al*** a house in London designed by Woolf Architects; ***25ac*** Fay and Roger Oates' house in Ledbury; ***25ar*** Dan and Claire Thorne's town house in Dorset designed by Sarah Featherstone; ***25bl*** a house in Devon designed by Anthony Hudson Architects; ***25br*** ph Christopher Drake/designed by McLean Quinlan Architects; ***26l*** ph James Merrell; ***26–27*** an apartment in London designed by Adjaye & Russell; ***27a*** ph James Merrell/Felix Bonnier's apartment in Paris; ***27b*** ph James Merrell/Amanda and Andrew Manning's apartment in Sydney designed by Stephen Varady Architecture; ***28*** an apartment in London designed by Ash Sakula Architects; ***29al*** Frédéric Méchiche's apartment in Paris; ***29cl*** James Merrell/a house in Sydney designed by Luigi Rosselli; ***29bl*** an apartment in Paris designed by Mark Guard Architects; ***29r*** Circus Architects; ***30al*** ph Christopher Drake/owners of La Cour Beaudeval Antiquities, Mireille and Jean Claude Lothon's house in Faverolles; ***30bl*** a house in London designed by Charles Rutherfoord; ***30br*** Circus Architects; ***31l*** ph Christopher Drake/Fay and Roger Oates' house in Ledbury; ***31r*** design Charles Rutherfoord; ***32–33 main*** ph Chris Everard/an apartment in New York designed by Gabellini Associates; ***34al*** ph Chris Everard; ***34bl&bc*** ph James Merrell; ***34–35*** ph Christopher Drake/a house designed by artist Angela A'Court, extension and alteration to rear of property by S.I. Robertson at 23 Architecture; ***35al&br*** ph James Merrell; ***36l*** ph James Merrell/a house in Sydney designed by Luigi Rosselli; ***36r*** ph Chris Tubbs/Giorgio and Ilaria Miani's Podere Buon Riposo in Val d'Orcia; ***37*** ph James Merrell/Khai Liew and Sue Kellet's house in Adelaide; ***38*** ph Simon Upton/Plain English; ***39*** ph Tom Leighton; ***40l*** Frédéric Méchiche's apartment in Paris; ***40c*** ph James Merrell/François Gilles and Dominique Lubar, IPL Interiors; ***40r*** ph Simon Upton/Lena Proudlock's home in Gloucestershire, which has since been restyled; ***41*** Frédéric Méchiche's apartment in Paris; ***42–43*** a house in Devon designed by Anthony Hudson Architects; ***43*** Dan and Claire Thorne's town house designed by Sarah Featherstone; ***44al*** ph James Merrell/the Ash house in London designed by Ash Sakula Architects; ***44ac*** ph James Merrell/François Gilles and Dominique Lubar, IPL Interiors; ***44ar*** a house in Devon designed by Anthony Hudson Architects; ***45al*** ph James Merrell/a house in London designed by François Gilles and Dominique Lubar, IPL Interiors; ***45ac*** ph Andrew Wood/a house at Jalan Berjaya, Singapore designed by Chan Soo Khian of SCDA Architects; ***45ar*** ph Christopher Drake/a house in Salisbury designed by Helen Ellery of The Plot London; ***46–47*** a house in Devon designed by Anthony Hudson Architects; ***48l*** ph James Merrell/François Gilles and Dominique Lubar, IPL Interiors; ***48c*** a loft in London designed by Robert Dye Associates; ***48r*** Linda Trahair's house in Bath; ***49*** ph James Merrell/an apartment in New York designed by Campion A. Platt, architect; ***50l*** ph Polly Wreford/Carol Reid's apartment in Paris; ***50br*** a house in London designed by Charles Rutherfoord; ***50–51a*** ph James Merrell/a terrace in London designed by Christophe Gollut; ***51a*** ph Christopher Drake/interior designer Carole Oulhen; ***51c*** ph Polly

Wreford/Liz Stirling's apartment in Paris; ***51b*** ph Christopher Drake/interior designer Carole Oulhen; ***52–53, 53l*** ph James Merrell/Frédéric Méchiche's house near Toulon; ***53r*** ph Christopher Drake/owners of La Cour Beaudeval Antiquities, Mireille and Jean Claude Lothon's house in Faverolles; ***54l*** Roger Oates Design; ***54c*** ph Christopher Drake/Valentina Albini's home in Milan; ***54r*** ph Christopher Drake/Nelly Guyot's house in Ramatuelle, France, styled by Nelly Guyot; ***55*** ph James Merrell/Khai Liew and Sue Kellett's house in Adelaide; ***56, 56–57a*** ph James Merrell/a house in Sydney designed by Interni Interior Design Consultancy; ***56–57b*** ph James Merrell/Linda Parham and David Slobam's apartment designed by Stephen Varady Architecture; ***57al*** Alan and Hepzibah's home in Sussex; ***57ar*** ph James Merrell/Amanda and Andrew Manning's apartment in Sydney designed by Stephen Varady Architecture; ***57b*** ph James Merrell/Interni Interior Design Consultancy; ***58ac*** ph Chris Everard/a house in London designed by Helen Ellery of The Plot London; ***59c&ar*** ph Chris Everard/New York City apartment designed by Marino & Giolito; ***58br*** ph Christopher Drake/Jonathan and Camilla Ross's house in London; ***59al*** ph Chris Everard/François Muracciole's apartment in Paris; ***59bl*** ph Chris Everard/Philippe Model's apartment in Paris; ***59ar*** ph Catherine Gratwicke/Agnès Emery's house in Brussels, tiles from Emery & Cie; ***59br*** ph James Merrell; ***60l*** ph Chris Everard/fashion designer Carla Saibene's home in Milan; ***60c*** ph Chris Everard; ***60r*** ph Erica Lennard/Nilaya Hermitage Hotel, Goa, India; ***61*** a mews house in London designed by McDowell & Benedetti; ***62l*** ph Chris Everard; ***62–63, 63l*** a loft in London designed by DAD Associates; ***63r*** ph James Merrell/Amy and Richard Sachs' apartment in New York designed by Vicente Wolf; ***64al*** ph Erica Lennard/La Chabaude, Apt, France; ***64ar*** ph Andrew Wood/Johanne Riss' house in Brussels; ***64bl*** ph Simon Upton; ***64br*** ph Ray Main/client's residence, East Hampton, New York, designed by ZG Design; ***64–65*** an apartment in Paris designed by Olivier Vidal AIA; ***66*** ph James Merrell/Vicente Wolf's apartment in New York; ***67*** Felix Bonnier's apartment in New York; ***68l*** a house in London designed by Woolf Architects; ***68r*** a house in London designed by Mark Guard Architects; ***68–69, 69a*** Charles Rutherfoord's house in London; ***69b*** a mews house in London designed by McDowell & Benedetti; ***70l*** ph Chris Everard/designer Karim Rashid's own apartment in New York; ***70ar*** ph James Merrell/Jan Staller's apartment in New York; ***70br*** ph Ray Main/Malin Iovino's apartment in London; ***71*** ph Chris Everard/Pemper and Rabiner home in New York, designed by David Khouri of Comma; ***72–73*** a house in London designed by Woolf Architects; ***73l*** ph Chris Everard/London apartment designed by architect Gavin Jackson; ***73cr*** ph James Merrell; ***74l*** ph Alan Williams/Jennifer and Geoffrey Symonds' apartment in New York designed by Jennifer Post Design; ***74br*** a loft in London designed by Robert Dye Associates; ***74–75*** Frédéric Méchiche's apartment in Paris; ***75c*** ph Chris Everard/an apartment in Paris designed by architect Paul Collier; ***75r*** ph James Merrell/an apartment in London designed by Charles Rutherfoord; ***75bl*** Charles Rutherfoord's house in London; ***76al*** ph James Merrell; ***76ar*** ph James Merrell/Keith Varty and Alan Cleaver's apartment in London designed by Jonathan Reed/Reed Creative Services Ltd; ***76b*** a house in London designed by Charles Rutherfoord; ***77a both*** ph Chris Everard/John Minshaw's house in London designed by John Minshaw; ***77bl*** ph Andrew Wood; ***77br*** ph Jan Baldwin/Alfredo Paredes and Brad Goldfarb's loft in Tribeca, New York designed by Michael Neumann Architecture; ***78al*** ph James Merrell/an apartment in New York designed by Pentagram; ***78ar*** ph James Merrell/Andrew Parr's house in Melbourne; ***78bl*** ph James Merrell/a house in Sydney designed by Interni Interior Design Consultancy; ***76bc*** ph James Merrell/a house in Noosaville designed by John Mainwaring; ***78br*** ph James Merrell/Interni Interior Design Consultancy; ***79*** Felix Bonnier's apartment in Paris; ***80a*** ph James Merrell/Andrew Parr's house in Melbourne; ***80c*** a house in Devon designed by Anthony Hudson Architects; ***80b*** ph James Merrell/Nicholas Larcombe and Caroline Solomon's house in Sydney; ***80–81*** a house in Devon designed by Anthony Hudson Architects; ***83al&ac*** a house in London designed by Charles Rutherfoord; ***83ar*** DAD Associates; ***83b*** ph Ray Main/central London apartment designed by Ben Kelly Design, 1999; ***84*** a house in London designed by Charles Rutherfoord; ***85a*** ph James Merrell/John F. Saladino's apartment in New York; ***85c*** ph Chris Everard/Nadav Kander and Nicole Verity's house; ***85b*** ph James Merrell; ***86*** ph James Merrell/a house in London designed by Charles Rutherfoord; ***86–87*** ph Chris Everard/John Minshaw's house in London designed by John Minshaw; ***87*** ph Debi Treloar; ***88al&ar*** ph Polly Wreford/Daniel Jasiak's apartment in Paris; ***88c*** ph James Merrell/Vincent Dané's house near Biarritz; ***88b*** ph Ray Main/Robert Callender and Elizabeth Ogilvie's studio in Fife designed by John C. Hope Architects; ***89*** Frédéric Méchiche's apartment in Paris; ***90*** Amanda Freedman's house in London; ***91al*** ph Debi Treloar; ***91ac&bl*** Amanda Freedman's house in London; ***91ar*** a loft in London designed by DAD Associates; ***91br*** a house in London designed by Mark Guard Architects; ***92a*** ph Ray Main/Jonathan Leitersdorf's apartment in New York designed by Jonathan Leitersdorf/Just Design Ltd; ***92bl*** ph James Merrell/an apartment in New York designed by Jacqueline Coumans, Le Decor Français with the help of Olivier Gelbsmann; ***92–93*** ph Alan Williams/Margot Feldman's house in New York designed by Patricia Seidman of Mullman Seidman Architects; ***93*** ph James Merrell/designer Charlotte Barnes; ***94c*** ph James Merrell; ***94r*** ph Jan Baldwin/Claire Haithwaite and Dean Maryon's home in Amsterdam; ***95*** ph James Merrell/François Gilles and Dominique Lubar, IPL Interiors; ***96*** ph Debi Treloar/artist David Hopkins' house in east London, designed by Yen-Yen Teh of Emulsion; ***96–97*** ph Jan Baldwin/the Meiré family home, designed by Marc Meiré; ***98a*** ph Debi Treloar/Cristine Tholstrup Hermansen and Helge Drenck's house in Copenhagen; ***98b*** ph Simon Upton; ***99al*** ph James Merrell/Sue and Andy's apartment in Blackheath; ***99ar&bl*** ph Simon Upton/interior designer JoAnn R. Barwick's house in Connecticut; ***100l*** ph James Merrell; ***100r*** ph Christopher Drake/Jane Churchill's house in London; ***101*** ph Christopher Drake/a house in Salisbury designed by Helen Ellery of The Plot London; ***102l*** ph James Merrell/Christophe Gollut's apartment in London; ***102c*** ph James Merrell/Shelly Washington's apartment in New York designed by Vicente Wolf; ***102r*** ph James Merrell/Linda Parham and David Slobam's apartment in Melbourne designed by Stephen Varady Architecture; ***103*** Felix Bonnier's apartment in Paris; ***104–105*** ph Alan Williams/private apartment in London designed by Hugh Broughton Architects; ***105l*** ph Polly Wreford/an apartment in New York designed by Belmont Freeman Architects; ***150cl*** ph Debi Treloar/Susan Cropper's family home in London—www.63hlg.com; ***105cr*** ph Andrew Wood/Phillip Low, New York; ***105r*** ph Chris Everard/designed by Filer & Cox, London; ***106*** ph Tom Leighton/a loft in London designed by Robert Dye Associates; ***107*** ph Chris Everard/designed by Filer & Cox, London; ***108–109*** ph Andrew Wood/media executive's house in Los Angeles, architect: Stephen Slan, builder: Ken Duran, furnishings: Russell Simpson, original architect: Carl Maston c. 1945; ***110l*** Richard Mabb and Kate Green's apartment in London; ***110–111*** an apartment in London designed by Robert Dye Associates; ***111*** Dan and Claire Thorne's town house in Dorset designed by Sarah Featherstone; ***112*** ph Alan Williams/private apartment in London designed by Hugh Broughton Architects; ***113a*** ph Chris Everard; ***113b*** floor by Dalsouple, First Floor; ***114*** ph James Merrell/Sue and Andy's apartment in Blackheath; ***115l & 116l*** ph Chris Everard/John Barman's Park Avenue apartment; ***116r*** ph James Merrell/an apartment in New York designed by Marino & Giolito; ***117*** ph Ellen O'Neill; ***118l*** ph Chris Everard/designed by Filer & Cox, London; ***118r*** ph Chris Everard/Richard Oyarzarbal's apartment in London designed by Jeff Kirby of Urban Research Laboratory; ***119*** Circus Architects, floor by First Floor; ***120*** © The Amtico Company—Honister Slate (HS83); ***121*** © The Amtico Company—European Slate (EA29); ***122*** an apartment in London designed by Emily Todhunter; ***123 both*** a house in London designed by Charles Rutherfoord; ***124–125*** an apartment in London designed by Emily Todhunter/rug by Garouste & Bonnetti; ***125l&r*** Roger Oates Design; ***126l*** ph Andrew Wood/Stephan Schulte's loft in London; ***126r*** ph James Merrell/rug by Woodnotes; ***127l*** ph Alan Williams/interior designer and managing director of the Société Yves Halard, Michelle Halard's own apartment in Paris; ***127ar*** ph James Merrell/Sarah Elson's house in London/rug by June Hilton; ***127br*** Roger Oates Design; ***128l*** ph James Merrell/an apartment in London designed by François Gilles and Dominique Lubar; ***128c*** an apartment in London designed by Emily Todhunter; ***128r*** ph James Merrell; ***128–129*** a house in London designed by Charles Rutherfoord; ***129*** ph James Merrell/Shelly Washington's apartment in New York designed by Vicente Wolf; ***130l*** an apartment in London designed by Ash Sakula Architects; ***130ar*** ph James Merrell; ***130br, 130–131*** a mews house in London designed by McDowell & Benedetti; ***132l*** ph Christopher Drake/designed by McLean Quinlan Architects; ***132r*** ph Chris Everard/Jo Warman—Interior Concepts; ***133*** ph Chris Everard/interior designer Ann Boyd's own apartment in London; ***134l*** ph Chris Everard/Jo Warman—Interior Concepts; ***134r&135*** floor by Helen Yardley; ***136l*** ph Tom Leighton/Keith Varty and Alan Cleaver's apartment in London designed by Jonathan Reed/Reed Creative Services Ltd; ***136–137*** Dan and Claire Thorne's town house designed by Claire Featherstone; ***137a*** ph Chris Everard/Charles Bateson's house in London; ***137b*** ph James Merrell; ***138–139*** Frédéric Méchiche's apartment in Paris; ***139ar*** ph Tom Leighton/Fay and Roger Oates' house in Ledbury; ***139br*** James Merrell/Andrew Parr's house in Melbourne; ***140*** Fay and Roger Oates' house in Ledbury; ***141a*** Frédéric Méchiche's apartment in Paris; ***141bl*** ph Ray Main/client's residence, East Hampton, New York, designed by ZG Design; ***141bc*** ph Debi Treloar/designed by Gordana Mandic of Buildburo; ***141br*** ph Christopher Drake; ***142l*** Linda Trahair's house in Bath; ***142ar*** ph Andrew Wood; ***142br*** ph Catherine Gratwicke; ***143*** ph Christopher Drake/William Yeoward and Colin Orchard's home in London; ***144a*** ph Christopher Drake/Fay and Roger Oates' house in Ledbury; ***144b*** ph James Merrell; ***144–145*** ph Simon Upton/Ann Boyd's apartment in London; ***145a*** Felix Bonnier's apartment in Paris; ***145b*** ph James Merrell/a house in London designed by François Gilles and Dominique Lubar, IPL Interiors; ***146l*** ph Andrew Wood/Roger Oates Design; ***146–147*** ph Simon Upton; ***147bl*** Charles Rutherfoord's house in London; ***147r*** Fay and Roger Oates' house in Ledbury; ***148l*** ph James Merrell; ***148r*** ph Andrew Wood/Roger and Suzy Black's apartment in London designed by Johnson Naylor; ***149*** ph James Merrell/rug by Woodnotes; ***150al*** ph James Merrell; ***150bl*** ph David Montgomery/Annabel Astor's house in London is full of furniture and accessories designed exclusively for her OKA Direct mail-order catalogue; ***151r*** ph Fritz von der Schulenburg/Alidad's apartment in London; ***152l*** ph Chris Everard/Hudson Street loft designed by Moneo Brock Studio; ***152c*** ph James Merrell; ***152r*** ph James Merrell/an apartment in London designed by François Gilles and Dominique Lubar, IPL Interiors; ***153*** ph Ray Main/David and Claudia Dorrell's apartment in London designed in conjunction with McDowell & Benedetti; ***154l*** ph Andrew Wood/Fay and Roger Oates' house in Ledbury; ***154–155, 155a&bl*** Fay and Roger Oates' house in Ledbury; ***155br*** ph James Merrell/a house in London designed by François Gilles and Dominique Lubar, IPL Interiors; ***156*** ph James Merrell/rug designed by Christine Vanderhurd; ***156–157*** an apartment in London designed by Ash Sakula Architects, rug by Christopher Farr; ***157a*** rug by Christopher Farr; ***157b*** ph James Merrell/Sarah Elson's house in London, rug by Christopher Farr; ***158al*** ph Andrew Wood/Norma Holland's house in London; ***158ar*** ph Andrew Wood/Chelsea loft apartment in New York, designed by The Moderns; ***158b & 158–159*** ph Andrew Wood/Jane Collins of Sixty 6 in Marylebone High Street, home in central London; ***159*** ph Alan Williams/interior designer John Barman's own apartment in New York; ***160*** ph James Merrell/Interni Interior Design Consultancy; ***161c*** ph James Merrell.